RESILIENCE
AND THE FUTURE OF
EVERYDAY LIFE

D1302126

JAMES H. LEE

Wasteland Press

www.wastelandpress.net
Shelbyville, KY USA

Resilience and the Future of Everyday Life
by James H. Lee

Second Printing – June 2012
ISBN: 978-1-60047-726-3

Cover design by Jason Boyett

Parts of this book have previously appeared online at *www.resilience-economics.com* and *www.wfs.org/blogs/james-lee*.

Chapter nine has been excepted for an article in *The Futurist*, titled "Hard at Work in the Jobless Future," March/April 2012.

Printed in the U.S.A.

0   1   2   3   4   5   6   7   8   9   10   11

More Advance Praise for
Resilience and the Future of Everyday Life

"Jim Lee provides us with a highly engaging and readable account that points out some very serious issues looming ahead. His persistent focus on solutions, with lots of examples and stories of the positive changes already taking place, gives us hope that we'll be up to the challenge. This is a great book for those who want to understand the challenges ahead and get a sense of what we as a society and they as individuals can do about them. His knack for storytelling makes this a difficult book to put down."

-Andy Hines, author of *ConsumerShift:*
How Changing Values are Reshaping the Consumer Landscape

"It's the end of the world as you know it and you should feel fine. But chances are, you'll only start feeling good once things make sense and you have a plan for your money and your life. Jim Lee's fantastic book will help you with both of these daunting tasks, explaining with tremendous clarity what is happening, and giving you insightful strategies for what to do next."

-Eric Garland, author of *Future, Inc.:*
How Businesses Can Anticipate and Profit from What's NEXT

"This practical and hopeful advice from a true futurist is a tapestry woven from the major threads of change that are morphing into the biggest shift in history... and provides a vision of how life could really be better."

-John Peterson, publisher and editor of FUTUREdition and author of *A Vision for 2012: Planning for Extraordinary Change*

"...takes the reader from the paralyzing darkness of the "gloom and doom" we hear so much about to the path of getting off our butts and into action, creating the reality we want for our children and their children – the epitome of resilience!"

-John Renesch, author of *The Great Growing Up: Being Responsible for Humanity's Future*

Contents

Introduction

It's the End of the World as We Know It
(And I Feel Fine) – R.E.M.

We still have not decided what to call the first decade of the new millennium. Maybe we should call it "The Big Zero." Columnist Paul Krugman likes this title, given that it was a decade "in which nothing good happened, and none of the optimistic things we were supposed to believe turned out to be true."[1]

Our friends on the other side of the Atlantic refer to the last decade as "The Noughties." Nought is their own fancy word for zero, but this also has the appeal of rhyming with "naughty," which describes bad behavior by people who should know better.

Another contender is "The Uh-Ohs." This somehow seems a bit more comforting. Mistakes have been made. Fingers have been pointed. It suggests a series of mishaps that could be cleaned up with a little attention and a sponge or a paper towel. "Uh-Oh" is also something that you would say *before* something bad happens. An "uh-oh" is not something that merits a panic attack, but it conveys a sense that there could be trouble soon. The dog is pulling the tablecloth, but the dishes have not fallen off the table yet.

We have seen many "uh-oh" moments over the past few years. The media isn't helping matters much. While listening to the evening news, the ongoing message is that things will only get worse. Environmentalists tell us that the planet is getting dangerously warmer. Analysts say that our government's credit rating has been downgraded. These are bigger problems than we have had in the past.

How will we know when things are going to get better? More importantly, what can we do to make things right sooner, rather than later?

Despite the unprecedented challenges, *I am an optimist.* Given a choice, I believe that people will generally do the right thing. Nobody really wants the world to fall apart. If we want better results for ourselves, though, we might need to do things a little differently.

There are many books about the future. For the most part, these fall into three broad categories: prophecies, predictions, and possibilities.

Prophecies typically come with a few significant drawbacks. They are usually vague and sometimes difficult to interpret. Prophecies serve as a warning to do things differently—or else! While prophecies are sometimes good at scaring people into behaving themselves, they offer little in the way of practical solutions.

In the modern age, we've attempted to turn prediction into a science. *Predictions* have a rigid and deterministic feel to them, as if there is no room for personal choice. With the passage of time, predictions can appear obvious, wrong, or at worst, obviously wrong. As a result, some books about the future should be printed with an expiration date.

This book is about *possibilities*, which is where most futurists prefer to work. It is a vision of what *could* happen if enough people choose to do things differently. You may find that some of these ideas fit your needs, while others don't seem to apply. That is perfectly fine. Go ahead, pick and choose what makes the most sense for you.

Introduction

Part One reviews the challenges ahead. We start with chapters on gloom and doom because that is what we hear about every day.

But this really isn't about the end of the world—it is about something else altogether. The Greek word *apocalypse,* when translated from its earliest meaning, is "to reveal; to uncover, to stand naked, exposed without artifice, clothing, or possessions."[2]

In this sense, we are experiencing a very real apocalypse. We are seeing governments, corporations, and cultures for what they are. We've entered into a time of full transparency and realize that we need a different set of values and systems. Fear-based responses are seldom beneficial for everyone involved. The best solutions make almost everyone feel a little better.

This raises the question of what might happen as we shift toward a more resilient way of life. A combination of self-reliance and social engagement may be helpful to navigate the challenges of the coming years. In the following pages, we will take a broad look at the many aspects of an emerging new culture.

While the problems are global, the solutions are personal. When enough people make similar choices, individual actions can lead to collective transformation. Some of these trends may seem obvious, while others only appear as weak signals.

Part Two is about the changes that we may see in our everyday lives. It includes inspirational stories and examples of how people are already solving problems at the grassroots level. We'll review existing key trends at length and include a section on weaker emerging trends at the end of each chapter. Finally, most chapters will have a resource guide for those who want to dig deeper.

If you are just looking for solutions, start with Part Two. If you are looking for trouble, be sure to read Part One. The first section of this book is about problems, the second is about possibilities.

Resilience and the Future of Everyday Life

How can we sustain ourselves in a world turned upside-down? We can do it by sharing, reconnecting locally, and focusing on what matters. It is basic, simple stuff that makes a difference.

Many people are in denial about global issues because they are either not listening or because it is so overwhelming that they don't know where to start. This book is about finding a place to begin.

Part One

Chapter 1

Gloom

When I was in college, there was a movie called *Weekend at Bernie's*. It is about two young insurance executives who are invited to their boss's beach house for the weekend, only to find that he is dead when they arrive.

There is a great party going on at Bernie's that night that nobody wants to stop, so they put a hat and a pair of sunglasses on Bernie, and then make excuses for why he is so mellow. "Oh, he's just meditating right now," they say. Or, "He's just a little stoned." No one seems to notice or care that the guy is dead, even as rigor mortis is setting in, so Bernie keeps getting propped up in different poses, and the party goes on...

We keep hearing interesting excuses about the economy, which this time is getting propped up by a man whose name is very similar to Bernie's. "It's a weak patch," he says. Or, "We're just seeing a slow growth recovery." Some of the excuses his friends are using are a bit more esoteric, like "reverse decoupling."

Meanwhile, the official statistics from the coroner, such as GDP growth and jobless claims, keep getting revised downward after an upbeat initial report is released. Nothing to see here...move along.

Longer term, we need to think about how to restructure in an economy with limited growth potential. It's not about "fixing" the economy with spending cuts and debt extensions. It's more about everyone learning how to do more with less. It's about simplying life. It's about rethinking net worth, not in terms of dollars but of friendships and family.

We are going to see lots of changes once we start to realize that Bernie is not going to be waking up from his nap anytime soon. Change is not all bad, particularly if we learn how to do things better than before.

This book starts with gloom and doom, because that's what we hear about on the news. There is a sense that things are broken and it will take a while to fix them. We were promised a new millenium, but so far, it has been a bit of bummer.

Collectively, it seems like we're all holding our breath, waiting for "something" to happen. We've been waiting for some defining moment to prove that the mess is over and that things will get better. There have been no such assurances.

Rather, it feels like we've been living in a state of suspended disbelief. The party at Bernie's goes on. The music has stopped, but people are still dancing.

Part of the problem is the bartender, who keeps serving shots of tequila long after closing time.

We've had one round of stimulus after the next. Government and Federal Reserve policies made in response to the collapse of the tech bubble and the Twin Towers led to the real estate bubble. The inflation of housing prices in turn allowed for a boom in lending and consumer spending. When the banks started to collapse in 2008, the only way to keep the system intact was through massive infusions of cash from the U.S. government, which in turn is funded by the Federal Reserve.

The current solution to the debt crisis seems to involve creating more debt on someone else's balance sheet. This is what happened when the government bailed out the banks. Cynical bloggers refer to this strategy as "extend and pretend." Clearly, this is not a sustainable system. Another round of tequila may delay a headache, but will not lead to sobriety.

David and Robert Weidemer comment on this situation in their book *Aftershock*:

> *Unlike at any other moment in our history, there is something fundamentally different going on this time. Even people who pay no attention to the stock market or the latest economic news say that they can just feel it in their gut. This is not merely a down market cycle, nor is it a typical recession...*
>
> *This is a bubble economy – one bubble built upon the next. Unfortunately, "a multi-bubble economy cannot be easily re-inflated."*[3]

A recent national poll by Rusmussen Reports shows that 72% of the country feels that we are on the wrong track; 20% feels that we are moving in the right direction.[4] Only 5% of voters say that Congress is doing a good job.[5] Meanwhile, President Obama continues to suffer low approval ratings, despite his best efforts.

What happens when the party is really over? Are we headed toward the worst hangover in history? These are some difficult issues to understand. Let's start with the questions that everyone is asking...

What Happened to the Middle Class?

According to the Census Bureau, 48% percent of all Americans are now living at low-income or poverty levels. The low-income threshold is currently at $45,000 per year income for a family of four. The poverty level is half this amount.

Paul Osterman, co-author of *Good Jobs America: Making Work Better for Everyone*, says that 20% of all working Americans have jobs that pay poverty-level wages. "The labor market is just not delivering for Americans what it should be delivering," Osterman says. "The weaknesses are on multiple dimensions—one is just the *quantity* of jobs and the other is the *quality* of jobs that do exist."[6]

Three decades ago, less than 30% of all jobs in the U.S. were considered low income. Today, that figure is over 40%.[7]

There are many reasons for the disappearance of the middle class. Pay cuts, forced reduction of work hours, and job losses all play a part. Making a living can be an expensive proposition – particularly for families with children. Housing and child-care costs can easily consume half of a family's income.[8]

"Safety net programs such as food stamps and tax credits kept poverty from rising even higher... but for many low-income families with work-related and medical expenses, they are considered too 'rich' to qualify," says Sheldon Danziger, a public policy professor at University of Michigan.[9]

Where Are the Jobs?

While the official statistics state that the unemployment rate is 8.5%, it is likely much higher. Close to seven people compete for every job opening, and many others have given up entirely. Three million Americans have stopped looking for a job, and eight million are working fewer hours than they would prefer. Once the discouraged and the underemployed are considered, the real level of unemployment in the U.S. is over 16%.[10]

Chronic unemployment is becoming more of a problem. People are not re-entering the workforce quickly. The average amount of time spent in unemployment has skyrocketed from an average of 15-20 weeks during previous recessions to over 40 weeks. Conservatives say this comes from the extension of unemployment benefits; liberals say that good jobs are hard to find.

Historically low interest rates, payroll tax cuts, bank bailouts and the rescue of the American auto industry have done little to improve the situation, although it is quite possible that unemployment would be much worse if these interventions had not been made.

Politicians have done everything they can to create new jobs. However, since consumer spending still accounts for two-thirds of the U.S. economy, there needs to be an increase in employment before people will feel confident again.

If these trends portend long-term changes in the labor market, we will need to rethink production and consumption in ways that are useful and realistic.

How Many People Rely on Transfer Payments?

Spending by the Federal government now accounts for 24% of GDP. [11] At the turn of the millenium it was closer to 18%. As the current bubble in government gets bigger, the economy grows more fragile and sensitive to changes in government spending.

There are now 8.5 million people receiving unemployment insurance and over 46 million receiving food stamps, according to the USDA.[12]

In 2009, the government paid Social Security benefits to 77 million people. Over half the recipients of Social Security rely on their monthy payments as their only source of reported income. [13]

Many more Americans are living close to the edge. A 2009 study by Metropolitan Life Insurance reported that half of all U.S. households do not have enough savings to cover their monthly expenses for two months in the event of a job loss. This is not just low-income families. Twenty-seven percent of households making over $100,000 per year have less than two months' worth of emergency savings.

Who Are the 1%, Anyway?

Occupy Wall Street and related protests have spread to over 200 cities worldwide. People are upset about growing income disparities between the wealthy and the poor.

The top 1% of income earners in the U.S. now account for 21% of total personal income.[14] This is the highest concentration of income that we've seen since the Great Depression.

In 1980, CEO pay equaled 42 times the median worker's pay. By 2010, CEO compensation had grown to 343 times the pay of a typical employee.[15]

Most people wouldn't be upset if the rich were simply getting richer, but there is a perception that some of them are benefitting at the expense of everyone else. Adjusted for inflation, the average U.S. household income fell by $3,719 over the past decade.[16]

Why Do We Have So Much Debt?

"The American Dream is now to get out of debt," says David Rosenberg, the chief economist and strategist at Gluskin Sheff.[17]

The Federal Reserve reports that total consumer debt in the U.S. is $2.4 trillion. This does *not* include home mortgages – this is credit card debt, car loans, students, and everything else. It equates to about $8,000 per every man, woman, and child.

The poorest twenty percent of American households had a (negative) net worth of -$13,800 in 2007, which declined to -$27,200 in 2009.[18]

Home equity was great while it lasted, but nearly 11 million homeowners are now "under water," meaning that they owe more in mortgage debt than their homes are worth. This works out to about 22.5% of all residential properties with mortgages, according to market research firm CoreLogic. [19]

In 2012, there will be a million foreclosures and a million-and-a-half bankruptcies.[20]

What Else Can We Worry About?

If you think you have problems paying your bills at the end of the month, consider the types of mail that the government is getting. Charles Hugh Smith writes the following spoof in his blog OfTwoMinds.com

> *Dear United States of America:*
>
> *We regret to inform you that your withdrawals exceeded your deposits last year by $1,600,000,000,000 ($1.6 trillion), including your "supplemental appropriations" spending.*
>
> *Your account does have an overdraft protection, and so bonds were sold to cover your $1.6 trillion overdraft. While we value your business, we feel obligated to remind you that this is the third year that your overdraft protection exceeded 10% of your gross national product (GDP), and it seems your account is on course to register yet another $1.6 trillion overdraft in fiscal year 2012.*

Currently, your overdraft account exceeds your GDP of $15 trillion. Quite frankly, we are worried that you have become dependent on extensive overdraft protection--a feature designed to tide the account holder over for a short period of time in near-term expectation of higher deposits or lower withdrawals--and that relying on large-scale overdraft borrowing to cover your basic expenses is now your standard operating procedure.

This violates the intent of the overdraft feature, and as a result we must seriously consider modifying the terms of the overdraft protection on your account. Current conditions enable us to provide this overdraft, but the feature was not designed to be permanent nor on this scale.

In order to give you sufficient time to bring your deposits and withdrawals back into alignment, we will maintain the current low-interest overdraft protection on your account through fiscal year 2012. Beyond that, however, please be aware that to maintain the integrity of the system, we will have to raise the rate of interest on your overdraft and scale back the size of the overdraft line of credit.

We regret informing you of these modifications, but the overdraft protection was not intended to be permanent nor near-infinite in scale.

Yours truly,

The Global Bond Market

What makes this funny (and scary) is that the numbers are quite real and accurate. According to the International Monetary Fund, the U.S. is currently positioned to have the largest budget shortfall in the world.

Gloom

As of 2011, the national budget deficit equalled 10.8% of our economic output. The total government debt in the U.S. has just exceeded 100% of GDP for the first time since WWII. The debt monster just swallowed our economy.

The total value of the U.S. federal debt works out to about $48,650 for every man, woman, and child in the country. This is equal to approximately $168,000 for every person working full-time. Everyone with a job will eventually pay Uncle Sam's mortgage.

Under President Obama, the U.S. debt increased by more than it has under all presidents combined, from George Washington through George H.W. Bush. The annual deficit has topped over a *trillion dollars a year* since Obama took office, and our government's debt is increasing at a rate of over $4 billion a day. [21]

Listen to the Occupy protestors and you will hear an easy solution – we can just raise taxes on the rich people. This would work, but only to a very limited extent. If Bill Gates handed over his entire fortune to cover deficit spending in the U.S., he could keep the government running for 15 days. Warren Buffett could keep things going for another 13 days. Put together, their resources would last less than a month. Keep in mind that the two richest men in America wouldn't be covering the total costs of running the government – simply covering the shortages in the current budget.[22]

Alternatively, we could tax all of our corporate profits. According to the U.S. Commerce Department, total corporate profits were about $1.4 trillion in 2010.[23] If all these profits were taxed an average rate of 100%, it would have been almost sufficient to cover the Federal budget deficit for 2011. This just covers one year's worth of spending deficits – the actual debt outstanding would still be unpaid.

So let's take this a step further. What if everyone emptied their piggy banks, wallets, cash in safe deposit boxes, checking, savings, money market funds, and certificates of deposit (excluding retirement accounts). What if we took all of this and used it to pay down the debt?

How much would this be? This number is known as the "money supply" and represents the total amount of cash in the U.S. economy – otherwise known as M2. According to the Federal Reserve, the total amount would be worth about $10 trillion.[24]   Let's imagine that we took all this cash and used it to pay down the debt.   Would that take care of the problem? It would be a good start, but our government would still owe $5 trillion.

Just How Far Away Is the Debt Bubble from Exploding?

Let's suppose that the interest rate on the U.S. debt goes up a mere 1%. Why worry?   On a $15 trillion debt, this works out to $150 billion per year. We can handle that, right?

A billion here, a billion there – soon we'll be talking about real money. What will $145 billion get you these days?   A few things – like the combined annual operating budgets of NASA ($19 billion), Homeland Security ($44 billion), the Department of Justice ($29 billion), the U.S. Postal Service ($6 billion), the Department of Energy ($30 billion) and the Department of the Interior ($12 billion).[25]

The U.S. is just three years away from having the same ratio of debt to GDP as Greece during the beginning of that country's debt crisis. "We're in deep trouble, and we're not alone," says Damon Vickers of Nine Points Asset Management.

One of the best-qualified and most outspoken critics of the U.S. budget situation is David Walker, the former U.S. Comptroller General under President Clinton and later, Bush.

Walker is something of a rock star among accountants. As head of the Government Accountability Office (GAO), he was the chief auditor of the country's finances. During 2005, while he was still in office, Walker embarked on a Fiscal Wake-up Tour, touring the country and publicizing his concerns about the unsustainability of the budget. In 2008, he was awarded the Gold Medal Award of Distinction by the American Institute of Certified Public Accountants – their highest recognition. Now he works with the Peter G. Peterson Foundation and continues to spread his message.

According to Walker, we are burying ourselves under a pile of debt that we cannot pay without mortgaging future generations: "Bottom line is we're not Greece, but we could end up with the same problems down the road."[26] He says that federal obligations and deficits far exceed the publicly stated levels of national debt. "We've got a couple of years to do something meaningful. If we don't, I think we'll have a crisis of confidence with regard to our own finances." [27]

Do You Hear the Demographic Time Bomb Ticking?

Walker worries about the $15 trillion dollar debt. However, what really keeps him awake at night are the unfunded liabilities of Social Security and Medicare.

According to the U.S. Census Bureau, the number of people at retirement age (65 and over) has increased from 8% of the population in 1990 to just under 13% of the population. Unfortunately for Social Security, new entrants and existing workers' pay the benefits of people already vested in the system – not unlike a pyramid scheme. This works so long as the bottom of the pyramid is growing faster than the top.

The Baby Boomers – 77 million of them – started to become eligible for Social Security in 2008, and some are now eligible for Medicare. This age group represents one quarter of the U.S. population. Beginning in 2011, Social Security started to burn through cash reserves; it now operates in a permanent cash flow deficit.

When there are more grandmothers than babies, the system is in trouble. Every government program that pays out lifetime benefits, from pensions to Social Security to Medicare, will feel the pressure. It is a pyramid scheme that simply will not work much longer.

As the population ages, there will be fewer workers putting money into the system to cover the costs of each Social Security beneficiary. According to the Social Security Administration, there were 3.4 workers per beneficiary in 1990. This ratio will likely decline to 2.1 workers per beneficiary within the next twenty years.

The problem with Medicare is that not only will there be more beneficiaries in the coming decades, but expenditures per person will increase as they grow older. While people between the ages of 65 to 74 only incur roughly $10,000 per year in average health costs, people over the age of 85 require over $25,000 in annual healthcare expenses.[28] This makes Medicare more of a problem than Social Security.

So on top of $15 trillion debt, there are also $38 trillion in unfunded future liabilities for Medicare, plus $7.9 trillion for shortfalls in Social Security. These are amounts that would be needed for investment *today* in order to meet payout demands in the future – in addition to expected future contributions. Without significant changes to the system, Walker estimates that the Social Security system will be bankrupt by 2038.

Once you consider the current underfunding of Social Security and Medicare, the total obligations of the U.S. government are closer to $200,000 per person, and over $500,000 per household.[29]

What Are the Alternatives?

We have gone past the point of fixing the problem using easy solutions. There are three basic policies that our government can pursue: print money, spend less, and tax more.

None of these are particularly good for long-term growth.

The government could print money. It could print as much money as it needed to pay the debt. Printing money creates inflation – it devalues the purchasing power of our currency and makes everything more expensive. Some economists refer to inflation as a "stealth tax."

The other options are equally unappealing. According to the Peter G. Peterson Foundation,

> *Stabilizing our national debt will not be easy. Using spending decreases alone to stabilize the debt, we would need to permanently cut the budget by 31%. Using revenue increases alone to stabilize the debt, we would need to permanently raise taxes by 44%.*[30]

The government could cut spending and send unemployment through the roof. Or it could raise taxes and create a wave of bankruptcies and foreclosures. Alternatively, it could systematically devalue the dollar through inflation.

Quite simply, there is no easy way out.

As our government obligations get bigger, the costs are going to rise over time. In a scenario of higher inflation, more taxes, and fewer benefits, what does one do?

It is a gloomy and bleak picture.

Clearly, we aren't going to solve this problem by doing the same things that we've been doing in the past. One definition of "insanity" is doing the same things over and over again and expecting different results. More government is not the solution. Businesses are part of the solution, but not the whole solution. The solution is really about the personal choices that we make. Individually, we can't change the government and most of us don't control businesses, but we can be responsible for ourselves and how we live.

We can all survive an economic crisis. I'm not at all worried about that. It's just gloom. Doom is another matter entirely.

Chapter 2

Doom

The Ocean is Dying

In 1997, Charles Moore was navigating his boat home from a yacht race in Hawaii. Looking for a more direct way home, he decided to cut through an area 1,000 miles off the coast of California called the North Pacific Ocean Gyre.

Sailors often avoid the area, given that it has sluggish currents and a lack of wind. Moore turned on his catamaran's motor, as he could not travel through this part of the ocean on sail power alone. He then spent the next week navigating through what he referred to as a "plastic soup" of abandoned grocery bags, disposable water bottles, and assorted refuse. Fifty years of accumulated refuse was slowly spinning in circles around the gyre – a giant drain without a downspout.

On land, plastic eventually deteriorates from exposure to sunlight. It can take centuries for plastic to decay into carbon dioxide and water. In seawater, the breakdown of plastics can take much longer – the ocean keeps the plastic cool while algae and other marine growth block ultraviolet rays.

The slow weathering of the elements grinds plastic into ever-smaller pieces. These eventually wash up onto the beach in pebbles known as "mermaid's tears." Or they can easily be mistaken for fish by marine birds, which consume the plastic and find themselves unable to digest it.

"It blew my mind," Moore said. "We are filling up the oceans with this confetti stuff and nobody cares."[31]

These are part of the synthetic seas – where natural life is slowly being displaced by plastic. A research report from the Algalita Foundation reveals that in the North Pacific Gyre, pieces of plastic outweigh zooplankton by a factor of 6 to 1. On the Midway Atoll, 40% of albatross chicks die. They are hungry, but their stomachs are full of trash. Each year, a million seabirds choke on plastic or get tangled in debris.[32]

The amount of plastic in the oceans has increased sharply in the decades since 1950. It has been estimated that the Great Pacific Garbage Patch (as the Gyre is now known) is now twice the size of Texas.

The oceans are experiencing other major threats as well.

Factory fishing is taking its toll. Computerized bottom trawlers drag miles of heavy nets across the ocean floors, gathering 130 tons of fish in a single catch – the maritime equivalent of clear-cutting. If it swims, it is caught. While trawlers are looking for bottom-dwelling fish such as sole, flounder, and halibut, they catch fish indiscriminately. Each year, more than 7 million tons of by-catch is netted and then thrown overboard. In the Gulf of Mexico, for every pound of shrimp harvested, four pounds of excess fish are captured and wasted.[33]

Half all coral reefs have disappeared worldwide, and populations of large predatory fish have declined by 90% in the past fifty years. The Newfoundland cod fisheries collapsed decades ago and still have not recovered. In Alaska, the king crab industry has seen a similar decline, along with the sardine industry off California. Fishermen are progressively moving down the food chain, harvesting smaller and smaller fish. Soon we will be eating bait.

Cynics should buy cans of tuna fish now – they will be much harder to find in ten years. Even now, restaurants on the Chesapeake Bay are serving "skate" (devilfish), which was once considered by-catch. In the Gulf of Mexico, some fishermen have moved from catching shrimp to harvesting jellyfish.

Primitive organisms are thriving on nutrients from fertilizers, sewage and other agricultural runoff. When combined with overfishing and a lack of competition, it creates a unique environment for more primitive organisms such as bacteria, seaweed, and algae to flourish.

Jeremy Jackson, a marine scientist with the Scripps Institute of Oceanography, refers to this as the "Rise of Slime." He is concerned that the evolutionary clock of the oceans is now moving in reverse. Jackson says that we are pushing back "to the dawn of evolution to a half-billion years ago, when the oceans were ruled by jellyfish and bacteria."[34]

Higher levels of carbon dioxide have made ocean waters 30% more acidic since the start of the Industrial Revolution, hurting shelled sea life (clams, corals, snails) while favoring species that evolved earlier in history.[35] Finally, surface waters have warmed slightly, about one degree over the past century, accelerating microbial growth.

This partly explains the recent resurgence of mass algal blooms around the world. Also known as "red tide," the *Karenia brevis* variety of marine algae is particularly troublesome. As a natural defense, it produces a neurotoxin that kills massive quantities of fish. The blooms create airborne particles that plague beach communities in Florida and other areas; the side effects for people include respiratory problems and memory loss.

Red tides used to occur about once a decade on the American east coast. Now they are occurring with greater frequency and severity.

Even non-toxic varieties of algae can cause problems if growth is unchecked. In some areas, dark algae blooms can easily be mistaken for oil spills. Dense swirls of algae can block light from reaching other plants. When algae dies, it is quickly consumed by bacteria, which absorbs oxygen in the process. The subsequent lack of oxygen causes snails, coral, oysters and other shellfish to suffocate. Fish flee the anaerobic water, leaving behind what could only be called a "dead zone" that is unable to sustain most marine life.

The largest dead zone in the world is in the Gulf of Mexico, just south of Louisiana. When viewed by satellite, it resembles a white cloud that swirls from the mouth of the Mississippi delta. It is now over 8,000 square miles in size – larger than the state of New Jersey. Other large dead zones exist where the Yangtze and the Pearl Rivers meet the ocean. According to a study that was published in *Science*, there are now over 400 dead zones in oceans worldwide – up from 49 in the 1960s. Most of these are in U.S. waters.

Just how bad can it get? "I can imagine an ocean that would be a continuous dead zone along all the coasts," says Jackson.[36]

There is good news here – these areas can sometimes be brought back to life when supplies of nitrogen are withheld. The Black Sea rebounded quickly when the collapse of the Soviet Union prevented local farms from buying fertilizer. Some progress has also been made along the Hudson River and San Francisco Bay. Like many other aspects of the environment, oceans take time to heal. The world may be better without us – for at least a little while.

Our inland water resources are experiencing a different type of threat.

Good to the Last Drop

Lake Mead is the largest freshwater reservoir in the United States. The Colorado River fills the lake from the north, while the lake powers the massive Hoover Dam to the south. Distribution pipes leading from Lake Mead eventually wind their way to homes throughout Arizona, Nevada, California, and even northern Mexico. They also irrigate over a million acres of farmland.

In addition to providing water and a source of power, Lake Mead is a popular recreation spot for boating, fishing, and waterskiing for residents in nearby Las Vegas. The city depends on the reservoir for ninety percent of its drinking water.

The lake is disappearing.

Circling the rim of Lake Mead is a one-hundred-foot tall natural "bathtub ring" of mineral deposits that marks its high-water level. After being nearly full in 1999, the lake dropped to 40% of capacity by 2010. The Scripps Institute of Oceanography estimates that there is a 50% chance that Lake Mead will go dry by 2021, if water demands continue to grow at expected rates. Researchers at Scripps have also predicted that there is a 50% chance that reservoir levels will be too low to allow hydroelectric power generation at the Hoover Dam by 2017.

The melting of ice in the Rockies has given a temporary boost of water flow to the Colorado River, but this was quickly consumed by immigration and development in the surrounding area. The result is that the lake is now running a huge water deficit each year.

"When expected changes due to global warming are included as well, currently scheduled depletions are simply not sustainable," wrote marine physicist Tim Barnett and analyst David Pierce in a peer-reviewed paper published in *Water Resources Research*.[37]

Eight studies completed from 1991 to 2007 have concluded that climate change will reduce the runoff from the Colorado River to Lake Mead by anywhere from 6% to 45% over the next half century.[38]

Like Lake Mead, the massive Ogallala Aquifer is also at risk. Unlike aboveground reservoirs, the Ogallala holds "fossil" water that is millions of years old and not easily replaced through ground sources. The Ogallala spans eights states between South Dakota and Texas and provides 30% of the groundwater used for irrigation in the U.S. Since it was discovered in the 1940s, the Ogallala Aquifer has been drawn down from an average of 240 feet to about 80 feet.[39] "We have drained enough water to half-fill Lake Erie," says David Brauer of the U.S. Agriculture Department.[40]

It is an invisible crisis. Because nobody really sees falling water tables or the over-pumping of aquifers, the problem doesn't attract as much attention.

Of all the water on our planet, only 2.5% is fresh water, and most of this is still frozen in glaciers and ice caps. Melt-offs of high altitude mountain ranges provide a one-time infusion of freshwater into depleted river systems. Unlike oil or other energy sources, water has no substitutes.

Lester Brown of the Earth Policy Institute says that "70% of world water use, including all the water diverted from rivers and pumped from underground, is used for irrigation, 20% is used by industry, and 10% goes to residences. Thus, if the world is facing a water shortage, it is also facing a food shortage.[41] Each ton of grain represents an investment of 1,000 tons of water. In many parts of the world, water equals food.

While 41% percent of the global population is experiencing some form of "water stress" (i.e., lack of clean water and adequate sanitation), population growth forecasts estimate that another three billion people will need water access by 2050.[42] In China alone, demand for water may grow 63% by 2030.[43] The Millennium Ecosystem Assessment report estimates that current levels of irrigation are 15-35% above sustainable amounts.   Even those with access to water today may not always have it in the future.

Even today, some of the largest rivers in the world are so stressed that they return little or no water to the ocean for months at a time. The Rio Grande, Yellow, Indus, Ganges, Amu Garya, and Nile rivers have all been tapped to a trickle. [44]

In Yemen, the water table under most of the country is falling by two meters per year. [45]   Chris Ward, an official at the World Bank, says, "Groundwater is being mined at such a rate that parts of the rural economy could disappear within a generation."[46]   They have been drilling wells over a mile deep – depths typically reserved for oil wells. Still, they are finding that they cannot discover new underground sources. The Yemeni government is considering bringing water from coastal desalinization plants or moving their capital city.

With the growing shortages in fresh water, combined with falling levels of soil fertility in parts of the globe, expect social unrest, economic disruption, and occasional outbreaks of violence.   As futurist Eric Garland says, "Whiskey is for drinking and water is for fighting over."

Dust to Dust

Soil fertility has been called the world's oldest environmental problem.

The Millennium Ecosystem Assessment forecasts that the global demand for food will increase by 70%-85% between 2000 and 2050. Food security is going to become a major issue without innovations in agriculture.

According to a study performed by John Crawford, a professor of sustainable agriculture at the University of Sydney, soil is being lost in

China 57 times faster than it is being replenished. In Europe, soil is disappearing at 17 times the replenishable rate, while in America the figure is 10.

David Tilman, Regents Professor of Ecology at the University of Minnesota says, "Global agriculture accounts for a third of all greenhouse gas emissions." Much of this comes from land clearing, which removes trees, creates erosion, and threatens wildlife. He says that agricultural greenhouse gas emissions could "double by 2050 if current trends in global food production continue."[47]

Monoculture farming, overgrazing, soil erosion, water shortages, and aggressive tilling have all contributed to loss of soil quality and quantity.

Running on Empty

The Ghawar oil field in Saudi Arabia is by far the largest of its kind in the world. Measuring 170 miles long by 19 miles wide, it is owned and operated by Saudi Aramco. It accounts for more than half all the oil produced in Saudi Arabia since it commenced production in 1951. So far, the field has produced roughly 60 billion barrels of oil. On a daily basis, it produces about 6% of the world's output.

It has been rumored that the Ghawar has entered into a phase of productive decline. While Aramco maintains a degree of secrecy regarding its oil reserves, the available evidence suggests that there may be trouble ahead.

In the 1960s, petroleum engineers started to inject water into the Ghawar to increase pressure. By 2004, 30% percent of the total fluids coming out of the reserve was water.[48] A daily infusion of more than seven million barrels of seawater per day was required to keep the oil pumping. There are now reports that suggest that the fluid coming out of the well is now more water than oil.

Doom

The usual life of an oil well is typically 50 years of production. This would make the Ghawar geriatric by most standards. In 2010, Aramco reported that it would soon be injecting carbon dioxide into the well,[49] an industry practice that is typically considered to be a last-ditch attempt to maintain production. Geosteering, 3-D seismic imaging, horizontal drilling, and maximum reservoir contact wells have all been deployed to siphon the last possible drop of oil from the field. The Ghawar appears to be coming up against geological limits, not technological ones. [50]

Of the remaining five "super-giant" oil fields around the world, all have officially entered into productive decline. China's Daquing field, Mexico's Cantarell, Russia's Samotlor, and Kuwait's Burgan have all passed their prime period of productivity.

Peak oil refers to the time when the rate of oil production starts an irreversible collapse due to depletion. This often occurs after the first 50% of available reserves have been extracted.

The heyday of oil discovery was in the 1960s, but since the 1980s, global oil consumption has outpaced discoveries. History has shown that peaks in oil discovery typically occur 30-40 years before peaks in production.

Many experts disagree on when peak oil will occur on a global level. We will only know when this has occurred with the benefit of adequate hindsight. Based on data from the U.S. Energy Information Administration, it appears that global crude oil production has stayed constant at about 74 million barrels a day since 2005.[51]

Production has already peaked in 54 out of the 65 largest oil-producing countries, including the United States.[52] The world is not running out of oil – just *cheap* oil. We are reaching limits in terms of how much can be produced each year.

The easy sources have been tapped out. When the first commercial rig was built in 1859, Edwin Drake only had to drill 70 feet before he found oil. Today, drillers off the coast of Brazil are developing reserves located 4 miles below the surface of the Atlantic Ocean.

Oil currently accounts for 41% of the world's fuel consumption and supplies 95% of the energy used for transportation.[53] Forecasts from the International Energy Agency estimate that overall energy demand will increase 33% during the next two decades.[54]

Gaps between energy production and global demand could create severe disruptions in the global economy. If this happens, we may see significant inflation in oil prices, along with other economic problems. Early in 2010, the U.S. Joint Forces Command issued the following warning:

> *By 2012, surplus oil production capacity could entirely disappear, and as early as 2015, the shortfall in output could reach nearly 10 million barrels per day…. A severe energy crunch is inevitable without a massive expansion of production and refining capacity. While it is difficult to predict precisely what economic, political, and strategic effects such a shortfall might produce, it surely would reduce the prospects for growth in both the developing and developed worlds. Such an economic slowdown would exacerbate other unresolved tensions, push fragile and failing states further down the path toward collapse, and perhaps have serious economic impact on both China and India. At best, it would lead to periods of harsh economic adjustment.*[55]

The Globe is Warming

Environmental activist Bill McKibben writes, "Global warming is no longer a philosophical threat, no longer a future threat, *no longer a threat at all.* It's our reality." McKibben continues, "The world hasn't ended, but the world as we know it has – even if you don't quite know it yet. We imagine we still live back on that planet, that the disturbances we see around us are the old random and freakish kind. But they're not. It's a different place. A different planet."[56]

The earth *is* changing. In summer of 2007, the thawing of the Arctic ice accelerated. By the end of the summer, 22% of the ice had melted from the year before, while the total amount of Arctic ice has diminished by close to 40% since the late 1960s. For the only time in human history, the fabled Northwest Passage above Canada is open and navigable. The first commercial ship to make the trip was the *MV Camilla Desgagnes*. While an icebreaker ship was available in event of blockage, one sailor reported to a Canadian news outlet that he "didn't see one cube of ice." [57]

The problem with melting ice caps is that they create a self-perpetuating cycle of global warming. Ice is far more reflective than water. Less ice means that more heat is absorbed by the oceans. This means more melting. Temperature change at the two poles seems to be happening quicker than in other parts of the planet. In 2008, *The Economist* reported that the West Antarctic was losing ice 75% faster than the previous decade.[58]

As the polar permafrost melts, it may release stored amounts of methane, a greenhouse gas that is twenty times more dangerous than carbon dioxide. It is estimated that almost ten trillion tons of this frozen gas are stored in the form of aboveground permafrost and beneath the earth's oceans. "The rate of increase in the Earth's atmosphere for methane is much faster than that for carbon dioxide," says Timothy Collete, a senior scientist of the Integrated Ocean Drilling Program (IODP).[59]

The effects of global warming are slowly being felt and recognized everywhere. Since the start of the industrial revolution, the average temperature of the planet has increased by an about one degree Celsius (1.8 degrees Fahrenheit).

Even the temperate zones may slowly become less temperate over the next few decades. Climate scientists warn about the advent of extreme weather, making the wet places wetter, and the dry places drier. The science is simple: Warm air holds much more water vapor than cold air. In warmer areas, this means increased rates of evaporation and drought. When that water vapor is transported to moist areas, it tends to come down in torrential downpours with high winds.

Extreme weather carries human and economic costs - the number of storm weather-related disturbances to the U.S. electrical grid has increased tenfold since 1992.[60] Meanwhile, disasters like the flooding of New Orleans and Haiti will become more commonplace.

Archaeologists are beginning to refer to the present as the Anthropocene era – the age when humanity changed the face of the planet more than any other species or geological event. James Lovelock, creator of the Gaia Hypothesis, said, "You could quite seriously look at climate change as a response of the system intended to get rid of an irritating species: us humans."[61]

In 2007, the Intergovernmental Panel on Climate Change (IPCC) estimated that average global temperatures will likely rise between 3.2 and 7.2 degrees Fahrenheit (1.8-4.0 degrees C) by the end of the century.[62] Mark Lynas examines what global warming really means in his book, *Six Degrees*. Lynas reviews data from the latest computer-driven simulations to make the following forecasts:

- An increase of **one additional degree** Celsius would cause a mega-drought in central and western U.S. states. The impact would be greater than the dustbowl of the 1930s. Sandstorms would engulf the entire region.   The southern coastal states would receive monsoon-like storms on a regular basis.

- **Two degrees:** The oceans' acidity level rises to a point where mussels, scallops and shellfish will no longer be able to survive. Middle East style temperatures sweep across Europe, and the southern Mediterranean countries lose a fifth of their rainfall.
- **Three degrees:**   The Amazon rain forest withers from drought. The southern and eastern coasts of the U.S. are battered by mega-storms.   Starvation replaces obesity as an epidemic. Tens of millions of Americans become "climate refugees."   At three degrees of warming, a third of all species could go extinct from severe habitat alteration.
- **Four degrees**: Antarctic ice sheets destabilize.   Sea levels rise dramatically, submerging coastal cities and island nations.
- **Five degrees:** Habitable zones become difficult to find and people must choose between regions prone to extremes of either extreme flooding or drought.
- **Six degrees:** The earth revisits the Cretaceous period, when oxygen levels plummeted to 15% and many organisms suffocated. At this point, life itself nearly dies.

Peak Everything?

Jared Diamond exhaustively studies why civilizations fail in his book *Collapse.*   His research indicates that there are five basic reasons:

- Environmental damage
- Climate change
- Hostile neighbors

- Loss of support of friendly neighbors/allies
- Failure of leaders to respond to growing problems

These issues tend to come in clusters, as one challenge leads to the next in a pattern that has repeated itself many times before. The initial underlying problem is usually a resource shortage, often resulting from environmental degradation. Civilizations originate and thrive where natural resources are abundant. Access to abundant food and useful materials allows for population growth, which eventually surpasses the supply of natural resources. Farming becomes intensified as lands are overgrazed and soil quality diminishes. Agricultural yields slowly decline. In an effort to secure resources, societies venture further away from their geographic center and into the territories of neighboring cultures. Supply chains weaken and the cost of importing commodities from far away becomes prohibitive.

Diamond's research indicates that societies facing resource shortages often come into conflict with competitors, either internally or externally. Climate change is sometimes the final event that pushes a culture beyond the tipping point by adding crop failures on top of existing resource shortages.

Along the way, denials of the problems at a political level prevent early interventions from occurring.

We are at a similar crossroads today. While most people are preoccupied with the state of the economy, they completely underestimate the environmental risks that we face over the following decades. This may be a confused ordering of priorities. As the late U.S. Senator Gaylord Nelson once said, "The economy is a wholly owned subsidiary of the environment, not the other way around."

The problems are deeply related. Understanding how to operate within the limits of our economic systems may teach us much about how to live within the bounds of our natural systems.

As far as civilizational challenges are concerned, gloom comes first and doom comes later. We are experiencing economic gloom in the here and now. My best guess is that we will have some resolution to the present gloom between 2015 and 2020.

It is not too late to avoid the worst aspects of doom. Planetary change seldom happens overnight; civilization has taken hundreds of years (or more) of resource depletion to bring us to the current point. Slower population and economic growth would provide a stabilizing influence, along with the potential cultural changes described throughout this book. Barring some cosmological event, I don't expect to see global doom happening any time before 2030 – and likely not before 2050.

Shift happens. People will need to change and adapt. Some of it will be easy; some of it will be difficult. Now, more than ever, we have access to the knowledge and tools that give us the foresight to make better choices. However, we cannot do it alone.

Community in Survival

Humanity can survive most changes – although perhaps not without some hardship. As the climate changes and natural resources are depleted, the pressures to survive at a basic level will touch our planet's poorest people first.

The most remarkable community that I ever encountered is on the other side of the globe. On the Tonle Sap Lake in Cambodia is a village of 5,000 people. Everyone lives on houseboats, and everything floats. The village migrates around the lake during the course of the year, based on water levels and the availability of fish. Nothing is wasted. If it can float, people find a use for it. The village is a relatively poor community by Western standards, but not necessarily so when compared to the surrounding agricultural area.

Nonetheless, it is encouraging to see how easily people could adapt to life on the water. In addition to the floating houses, there is a temple, several schools, a library, and even a basketball court.

People there are largely self-sufficient. Just about everyone has a floating fish pen for aquaculture. They also have floating pens for hogs and chickens. Crocodiles are raised for food and leather. (There were no signs or disclosures like "please don't pet the crocodiles"- just a wide-open hold in the middle of a boat filled with 20 or so crocs.)

Children get around from one house to another by paddling small boats and washtubs. Overall, they seemed well accustomed to life on the water, and all of them learned how to fish at an early age.

Comforts are sparse, but people have the basics. Most homes have floating stacks of firewood for heating and cooking. Some have a television antenna. Electronics are powered by batteries and generators.

A good friend of mine spent a year living in the village while studying the wild crane population. She said that living on the lake made the evenings cool and that sleeping on the water was comfortable. If people didn't like their neighbors or "needed space," they would just move their house somewhere else. Sometimes it was a little hard finding people, because the "buildings" kept moving. There were no streets, and no fixed addresses.

I left the village with a feeling of hope and respect. It proved to me that people could adapt to just about anything. If the waters rise and flood the great cities of this planet, the people of the Tonle Sap Lake will just keep paddling away.

Chapter 3

Toward a New Culture

In childhood, we played games that rewarded thinking big and dreaming big. We built real estate empires playing *Monopoly* and global empires playing *Risk*. *The Game of Life* was a board game intended to recreate the passages of life, complete with jobs, marriage, and mortgages. In practice, the game involved filling a plastic car with peg-like children and racing to the "Day of Reckoning" with the largest bank account possible.

We've grown up. Today, we have bridges to nowhere and streets to everywhere. We have too much stuff to put in our too big houses.

How did we all get here?

Writer and futurist John Renesch looks back at the last few decades, and notes that "perhaps it was a reaction to the 1929 stock market crash and the Great Depression in which people of my parents' generation suffered so much, but a 'chicken in every pot' gave way to 'two cars in every garage'."[63]

The mass consumer culture gained momentum at the end of WWII. After years of poverty during the Great Depression, followed by shortages and rationing during the war, people were ready to focus on starting new families and new lives. The large-scale industrialization of resources built for the war created an infrastructure for a new consumer economy. It took a few visionaries to understand how to make that transition possible.

William Levitt was one of the first to apply mass-manufacturing techniques to the housing industry. While most homebuilders were building three to four houses per year, Levitt was building thirty to forty houses a *day*. Levittown became one of the first great suburbs of the post-war generation. The first model homes had two bedrooms and measured 750 square feet. These homes also lacked many of the amenities that we would expect. There was no garage or basement and the second floor was unfinished.

Levitt chose prefabricated parts whenever possible, including metal cabinetry. Buyers seldom had any money left over for appliances, so Levitt purchased refrigerators and ovens directly from General Electric and sold them as part of the house. The marketing strategy was to build every home for less than ten thousand dollars, so that buyers could get federally backed loans.

Levittown became a bustling community of seventeen-thousand homes. So many children were born in the community during the 1940s and 1950s that it gained the nickname "Fertile Acres."[64]

Consumer culture was in full bloom. For the first time, corporations were changing not only what we chose to buy, but also our reasons for buying. Victor Lebeau, a marketing consultant and retail analyst, wrote the following in a 1955 issue of the *Journal of Retailing*:

> *Our enormously productive economy demands that we make consumption our way of life, that we convert the buying and use of goods into rituals, that we seek our spiritual satisfactions, our ego satisfactions, in consumption. The measure of social status, of social acceptance, of prestige, is now to be found in our consumptive patterns. The very meaning and significance of our lives today expressed in consumptive*

terms. The greater the pressures upon the individual to conform to safe and accepted social standards, the more does he tend to express his aspirations and his individuality in terms of what he wears, drives, eats – his home, his car, his pattern of food serving, his hobbies.

These commodities and services must be offered to the consumer with a special urgency. We require not only "forced draft" consumption, but "expensive" consumption as well. We need things consumed, burned up replaced, and discarded at an ever-accelerating rate.

The chairman of President Eisenhower's Council of Economic Advisors, Arthur Burns, later made this national policy when he stated, "The American economy's ultimate purpose is to produce more consumer goods." The best way to achieve this was through mass manufacturing of products that would quickly become obsolete – either through planned defect or through intentional disposability.

There is also something called perceived obsolescence, which encourages the replacement of perfectly useable items in order to stay in fashion and to maintain social status. The average American is now exposed to 3,000 advertisements in a day – all for the purpose of selling more stuff.

The first homes constructed after WWII could no longer hold the rapidly expanding American dream. The public needed more and bigger houses faster than ever before. By 2009, the average new home had 2,700 square feet of space – more than three times larger than the homes that Levitt originally built. According to U.S. Census Data, new homes grew by over 800 square feet of living space in the two decades leading to 2007.

Even our current monuments to mass consumption are inadequate to fulfill our needs to keep and store our stuff. Despite ever-larger houses with attics, garages, and sheds, the self-storage industry has been booming in the U.S. and is now a $22 billion business. There is now seven square feet of self-storage space for every man, woman and child in America.

These facilities could best be described as "condominiums for stuff" – kept out of sight (and out of mind) in safe, climate-controlled conditions. While the industry has been around for three or four decades, over a *billion* square feet of rental space was built between 1998 and 2005.

Aside from problems of storage, other problems arise from a culture of mass consumption.

The Missing Link

The "Bigger, Faster, More" syndrome is the common cause of both economic and environmental instability. This addiction to growth has perpetuated unsustainable levels of debt and resource consumption. As with other addictions, larger doses are continually needed to keep the good feelings going.

Psychologists have observed that addicts tend to stay at the level of maturity that they reached at onset of their addiction. In this way, a 30-year old addict can have the relative maturity of someone twelve years younger. Part of this comes from a denial that anything else matters.

Duane Elgin and John Renesch are two futurists who have independently met with audiences around the world and surveyed them about the current level of development for the human race – child, teenager, adult, or elder. The near-universal response was that we are now at an adolescent stage of maturity.

What does this say about us? According to James Hollis, former Executive Director of the Jung Educational Center for Houston:

> *The characteristics of an adolescent culture are: poor impulse control, short-term memory and ignorance of history, addiction to novelty and new sensations, addiction to escalating intensity of sensation, susceptibility to the lethargic summons of drugs and intoxicants... and mostly flight from independence and responsibility.*[65]

The current boom/bust cycles of our economy are similar to the dramatic mood swings of a teenager. Drama and self-absorption are commonplace, along with a desire for freedom without accountability. Teenagers tend focus on self-image and personal development; they have not fully developed the skills of partnership or caring for others.

Many mainstream economists, like teenagers, believe that the party should never stop. An astoundingly large portion of our economic system relies on perpetual growth. Growth is necessary to keep pensions going and shareholders happy. We measure economic progress not by the fulfillment of basic needs, but by how much more we are making and consuming relative to last year. A decline in growth means that CEOs get fired by their board and presidents get voted out by the public.

Robert Samuelson writes in *Newsweek*, "We Americans are progress junkies. We think that today should be better than yesterday, and that tomorrow should be better than today."[66]

The addiction to growth and the need to consume have created spiraling levels of personal debt that will collapse if the growth engine stops moving. Forty percent of American families annually spend more than they earn, and sixty percent carry balances on their credit cards on a monthly basis.[67] Furthermore, close to thirty percent of home mortgages are now underwater (meaning that the remaining balance on the mortgage is greater than the value of the house).[68]

A collapse in growth leads to a collapse in the ability to manage debt. Lost jobs lead to missing mortgage payments. Defaults accumulate to create bank closures, and sometimes bank closures lead to spiraling national debt. It is a chain reaction that is well documented in Carmen Reinhart and Kenneth Rogoff's sarcastically titled book *This Time Is Different.*

The more things change, the more they stay the same...

The real problem here is the problem of managing expectations of exponential growth within the confinements of a closed system. Everywhere along the way, our culture of mass consumption is bumping up against limits – both financial and ecological. In *The Story of Stuff,* Annie Leonard comments,

> *One third of the planet's natural resource space has been consumed. We have less than 4% of our original forests, and 40% of our waterways are now undrinkable. In the U.S., we have 5% of the world's population, but are using 30% of the world's resources and creating 30% of the world's waste. 75% of global fisheries are fished at or beyond capacity... we release 4 billion pounds of toxic chemicals per year.*

Ecologists estimate that our global population was sustainable until the early 1980s. At that point, we started to break open the environmental piggy bank of accumulated resources. We continue to consume the planet at a rate that exceeds its annual level of resource production. This leads to the question of what would be a reasonably sustainable level of global consumption. Herman Daly, an ecological economist from the University of Maryland, has put together the following rules to help define the limits of global resource use:

- Renewable resources (including soil, water, forests, and fish) cannot be consumed at a rate faster than their rate of growth or regeneration.
- Nonrenewable resources (such as petroleum, minerals, and fossil groundwater) cannot be consumed faster than the rate at which renewable resources can be substituted for them.
- Pollutants cannot be emitted at a rate greater than they can be recycled, absorbed, or rendered harmless.[69]

Using this analysis reveals a serious problem – the world's population needs one and one-half Earths to sustain its current level of consumption. If everyone lived an American lifestyle, we would need over five Earths.[70] We only have *one.*

Within both financial and environmental frameworks, spending beyond income is unsustainable for any length of time. The eventual consequences may be bankruptcy in both instances. While the risks of financial bankruptcy are readily visible on personal and national levels, we may not be fully aware of the risks of ecological bankruptcy for another few decades.

Both crises stem from the same root cause – excessive consumption within the context of limited resources. If we can sustainably solve the problems of financial gloom, we have a decent chance of beating environmental doom.

There are three ways out of the current situation: time, technology, and social transformation. Of these three paths, social transformation may provide the most immediate, yet enduring, solutions.

Time

One possible solution to the current economic crisis would be waiting until the next cyclical recovery. This could work if all we need is a rebound in consumer spending to rejuvenate the economy and balance the budget deficit. Obviously, the economy is more complicated than that, but consumer spending fuels seventy percent of economic activity in the U.S. Using this premise, we simply need to figure out when consumers will start to feel comfortable spending again.

In forecasting the future, many practitioners favor trend analysis – which is simply an extrapolation of current patterns into the future. While trend analysis can provide useful indicators of direction and momentum, it often fails at identifying key turning points. Forecasters are then surprised when their predictions turn out to be inaccurate. Proponents of cycle theory feel a major advantage of using cycles is that they can identify key turning points years in advance.

The best-researched long-term economic cycle is the Kondratieff wave. Discovered in the 1920s by the   Russian economist Nicolai Kondratieff and described in his book *The Major Economic Cycles,* a complete cycle typically lasts between fifty to sixty years and consists of alternating periods of fast and relatively slow economic growth. Each major cycle is initiated by a technological revolution, such as the steam engine, railways, electrical engineering, petrochemicals, or digital computing.

Kondratieff divides his cycle into four seasons. The *spring* is a period of stabilization and recovery, followed by a prosperous *summer* of prosperity. *Fall* is a period of maturing growth, while *winter* is a period of depression, unrest, and international conflict.

Following this cycle, the current winter period may continue into 2017.

There are a few other forecasting tools worth exploring, as well. Demographic data sometimes provides the most stable and accurate basis for long-term forecasts. Unfortunately, if we just look at population statistics, the future continues to look dim for the next few years.

Harry Dent is one strategist who made a name for himself by looking at the numbers of people entering and leaving the peak spending periods of their life – starting in their late twenties and ending in their late forties.

Using this approach, Dent successfully predicted the 1990s boom and the subsequent market collapse in 2007. Just as the Baby Boom generation fueled the rapid expansion of the consumer economy in the past few decades, he expects their children (sometimes known as Echo-Boomers, or the Millennial generation) to eventually fill in the spending gap - but not until the early 2020s.[71] Before we reach that point, however, he warns that unemployment will reach the 15% level and that the stock market may decline as much as seventy percent. So much for optimism...

There are multi-generational cycles worth considering, as well. William Strauss and Neil Howe focus on a measurement of time called the *saeculum* – roughly equal to one human lifespan and divided into four turnings. Each season in this cycle is marked by the birth of a new generation. These eighty-five year cycles frequently end in twenty-year periods of turmoil. Past saecular winters resulted in the American Revolution, the Civil War, and WWII.[72] Based on this analysis, the current period of crisis period could run until 2020-2022.

As a cure for economic gloom, time is on our side. We can get through the current period of economic malaise, as countries have done many times in the past. However, time is no cure for a weakening environment.

Eamon O'Hara, a policy advisor for Ireland, writes, "Obviously, the first thing we need to do is act, and act fast. Every day we wait... between 100-150 plant and animal species become extinct; 70,000 hectares of rainforest is destroyed, and another 150m tonnes of C02 is released into the atmosphere."[73]

As long as we continue to place stress on our planet, it will not have time to heal and repair itself.

Technology

Improvements in technology can stimulate the economy in any number of ways. Technology can increase economic output by creating higher levels of productivity. High productivity often makes related products and services more affordable. Finally, new technologies can generate consumer demand by introducing new "must have" products to the marketplace.

Technology plays an important role in every aspect of our everyday lives. It is worth noting that some of our most essential technologies enable the creation and use of energy (electric, heat, combustion). Energy is at the intersection of both environment and economic concerns. As such, it deserves further attention.

If we consider the problem of managing the global demands for energy, we can identify up to four broad classes of solutions: generation, transmission, storage, and usage.

Cleaner power may someday come from existing coal, oil, and gas plants by using carbon capture storage technologies. One promising technique burns coals and purified oxygen into an easily harvestable metal oxide, which is then captured and stored. Other carbon capture technologies pump waste carbon dioxide back into the earth, rather than the atmosphere.

Renewable biofuels may eventually replace traditional fossil fuels in transportation. The most productive biofuel source is algae, which creates about 15 times more useable fuel per acre than corn. Algae grows year-round in temperatures ranging from below freezing to 158 degrees Fahrenheit. In theory, the U.S. could produce enough algae-based biofuels to meet all of its transportation needs without adding new carbon dioxide to the atmosphere. Furthermore, algae can feed on waste carbon dioxide from coal plants and nutrients from agricultural manure.

Electrical energy could be generated in any number of novel ways in the coming decades. Space-based solar power can create sustained energy twenty-four hours a day, free of cloud cover or atmospheric effects. Light from giant orbiting solar panels could be harvested by satellite and safely beamed back to the planet via microwave transmission. Pacific Gas & Electric recently signed a power purchase agreement with a company that hopes to do this. A thousand megawatt plant could produce energy to power over a million U.S. homes.

Search giant Google is putting hundreds of millions of dollars into their "RE<C" initiative, which seeks to make renewable energy cheaper than coal. After renewable energy costs reach this level, there will be no economic reason to continue using non-sustainable sources.

As far as power transmission is concerned, many organizations are working on smart grids that better balance production with consumption. Eventually, room-temperature superconductors such as graphene could transport electricity over long distances without suffering significant heat loss. Other superconducting materials can enable the manufacture of hyper-efficient motors and wind turbines.

Dozens of companies are working on finding better electric storage technologies. Lithium-air batteries could potentially be significantly more efficient than current lithium-ion batteries and weigh a fraction as much. Better batteries could extend the driving range of current electric vehicles by a factor of ten and reduce the demand for fossil fuels significantly, provided that the energy for those vehicles is produced from non-fossil sources.

While energy-efficient appliances, hybrid vehicles, and compact fluorescent bulbs have already had an impact on energy consumption, many of the game-changing technologies are still years away from mass commercialization. Space-based solar power is still a "pie in the sky." We will not see mass commercialization of algae farms for at least another decade, and superconductive power transmission cables are only in a few locations worldwide.

There are other reasons technological fixes alone cannot be relied upon to save us from doom.

In the mid-1860s English economist William Stanley Jevons noticed that increased efficiency in the use of coal led to higher levels of coal consumption throughout industry. Improved designs for steam engines made coal a far more cost-effective power source, leading to a much broader usage of both the Watt steam engine and of coal. Jevons wrote in *The Coal Question,* "It is a confusion of ideas to suppose that the economical use of fuel is equivalent to diminished consumption. The very contrary is the truth."

The same principle explains why adding more lanes to an existing highway will only provide temporary relief of traffic congestion. Greater ease of transit encourages longer commutes and more traffic.

Cheaper, more efficient energy could also have other unseen spillover effects, including a higher consumption of other natural resources due to lowered costs of extraction, processing, and transportation.

Finally, innovative technological solutions tend to create financial bubbles. As enthusiasm for emerging technology grows, over-investment often follows, creating a misallocation of resources and an eventual collapse in asset prices. We have seen this time and time again.

So technology provides a fix -- albeit only a temporary one. Over time, though, even the wonders of technology may be inadequate to keep up with our hopes for "bigger, faster, more."

Transformation

The third way to move beyond the current crisis is through social transformation. Grassroots change will need to happen, because governments no longer have the resources to direct policies as they have done in the past. Furthermore, political leaders are finding that each new intervention is becoming progressively less effective at maintaining "business as usual."

There are limits to the benefits of government programs such as tax cuts and spending for new jobs, according to futurist Andy Hines. "We're trying to stimulate a caffeine junkie on their twelfth cup of coffee with a thirteenth cup. It's no longer having the intended effect," Hines notes. "The Great Recession has enabled the postmodern values shift toward less emphasis on material goods consumption." As a result, there are signs pointing toward a growing sense of "enoughness," and people are now searching for ways to "get off the treadmill — the constant pressure to keep consuming. They are stepping back, re-prioritizing, and looking for ways to do with less. "[74]

This is a time for us to become more resourceful than ever before.

So the question is – what does a thriving economy look like if it is not growing? If we are not consumers, then what are we? If the answer to our problems is no longer having more of everything, how will we find the solution?

At some point, our emerging new culture may recognize that we already have enough. Enough is truly *enough* when we recognize that there are already adequate amounts of most things to go around. It is really a matter of distribution – getting things *where* they are needed in an effective manner.

We are at the edge of a potentially remarkable series of transformations as we move from our "old culture" to a "new culture" that is balanced and sustainable. These shifts can change every aspect of our lives. Here is a list of possibilities:

- Economic relationships change from ownership to access. People learn how to share better than ever before.
- Less is more. People consciously simplify their lives to obtain greater personal flexibility and freedom.
- Daily household activities shift from consumption to production. Families learn how to become more self-sustaining through small gardens, do-it-yourself activities, and home-based businesses.
- Nothing is wasted. The new motto is "reuse, recycle, and repurpose." Every unit of waste becomes a resource for something else.
- Found and extended families replace government programs as providers of social security.
- A patchwork quilt of work opportunities replaces the traditional career ladder. Some industries see the gradual disappearance of the physical workplace and the workweek. Employers value results above the amount of time spent working and in meetings.
- Physical goods and services become increasingly local, while the information economy becomes more global.
- Wealth is more frequently measured in terms of reputation.
- Healthcare moves from a model of sustaining life at any cost toward providing quality of life.

These social transformations could all be a part of our collective growing up that enables us to take care of our families and ourselves without adding further financial and environmental stress. These and other cultural changes will likely not occur out of collective personal choice or preference, but out of necessity.

We may become somewhat poorer in terms of material goods, but we may also become considerably richer in terms of well-being and experience. Also, by working together and sharing resources within our communities, we may be able to rebuild a sense of collective trust that has collapsed over the past years.

The philosopher Immanuel Kant wrote, "Enlightenment is man's emergence from his self-incurred immaturity." The twin challenges of gloom and doom may give us the push that we need towards a wiser, more mature culture.

Social change can take many different forms. It is difficult to predict the exact outcome, but it is clear that momentum is gathering to provide transformational changes to our culture over the next decade – many of them positive.

Paul Hawken sums it up beautifully in his book *Blessed Unrest:* "This is the largest social movement in all of human history... coherent, organic, self-organized congregations involving tens of millions of people dedicated to change... If you look at the science that describes what is happening on earth today and aren't pessimistic, you don't have the correct data. If you meet the people in this unnamed movement and aren't optimistic, you haven't got a heart."

Part Two

Chapter 4

Make Me a Revolution

The last time the world came to an end was in the 1970s. The stock market collapsed, oil prices went through the roof, and we were closing out an unpopular war. Sound familiar?

There was a feeling that things were coming apart at the seams. My family responded by moving into a tiny town with a post office, a general store, and a gas station. A church and a restaurant were on the far side of the railroad tracks. Our mailing address was P.O. Box 4. What we lacked in neighbors, we more than overcompensated for in open space.

My parents started buying things in bulk to save money and stock up for shortages. Our kitchen couldn't hold everything, so Dad put up storage racks in the garage.

They also learned how to do stuff. Mom had always been good with a sewing machine and made some of her own clothing from patterns. Dad started to grow Concord grapes in the backyard, which he boiled to make jelly and then sealed with wax in glass Ball jars. For a brief but memorable period, Dad experimented with cutting the hair of his three children in the machine shop, using a kit that he bought at Sears Roebuck. He also installed an enormous cast-iron Franklin stove in the family room to provide plenty of heat in the winter. The flat top of the stove was warm enough to boil water for tea and to make soup.

My father was always a maker of things. There was comfort in knowing that if we ever really needed anything, he could probably make it, *MacGyver*-style, out of aluminum and plywood. It started out when he got married as a graduate student and built a crib for his first daughter using hand tools. Later, he built the furnishings for his machine shop and his own hi-fi system. His tinkering habit then progressed to projects of increasing complexity. Everything that he made looked and felt like a prototype for something that hadn't been invented yet. When I was in college, he built a metal detector that was so sensitive that it could detect the iron in your body at a distance of several feet. Later on, he developed a "perpetual motion" machine that would start spinning as soon as you held it. There were no solar panels or batteries involved in the design – it ran on the convection currents generated by the heat of your hands.

So (with my Dad's help) I was the kid with the best science projects and the fastest pinewood derby car. It was a good environment for thinking and experimenting.

We had a Commodore VIC -20, with a cassette-tape storage drive and 2 kilobytes of programmable memory (which we soon expanded to 7 kilobytes). The memory constraints required tight coding in BASIC, and the only monitor we had was our Trinitron TV. The VIC-20 itself looked like nothing more than a chunky keyboard.

Instead of buying cartridge games for our computer, my folks had the foresight to have me do my own programming when I was 12. For a few years in the 80s, magazines such as *COMPUTE!'s Gazette* would print lines of code in the back pages, so I started copying and modifying programs. Pretty soon I was writing my own Pac-Man knock-offs. Most of my games were not that memorable. Quite frankly, they reminded me of B-movies. A certain amount of imagination was required to make them better than they actually were. But it also put me on the path toward thinking systematically and being able to program my own stuff.

DIY Is Good for the Soul

As the world comes to an end again, it seems that the interest in learning how to do things is experiencing a revival. DIY (doing it yourself) is rapidly becoming a trend that is going mainstream. It may be out of some basic survival instinct, or possibly the feeling that something made or handcrafted is more "real" or genuine than anything mass-produced. There comes a satisfaction in knowing that if something is broken or lost, you can rebuild it from scratch. The knowledge and experience is yours to keep. Better yet, there is no longer any need be satisfied with what you can buy at Wal-Mart. Everything that can be made can be built according to your specifications – provided that you have access to the appropriate plans, expertise, tools, and materials.

In addition to the comforts of self-sufficiency and the gratification of creating customized items not available commercially, there are many other reasons the maker and crafting movements are here to stay for a while.

Mark Frauenfelder is the editor-in-chief of *Make* magazine and founder of the popular Boingboing blog. He never really thought of himself as being a "hands-on" type of person – his background was in writing and computers. But over the past few years, Frauenfelder has made a shift toward hacking into the real world. "You can make your world a little less confounding by sewing your own clothes, raising chickens, growing vegetables, teaching your children, and doing other activities that put you in touch with the processes of life. In addition, the things you make reflect your personality and have a special meaning. You share a connection with them every time you use them, and you appreciate them in a different way than you do store-bought things."[75]

Frauenfelder also says, "It connected me with my kids and made me more appreciative of what is around me."[76]

The New Cottage Industry

Access to knowledge, ideas, and material has never been easier than it is today.   An entire cottage industry has sprouted around the DIY movement. The Internet has enabled anyone to become an expert on just about anything within the course of a few hours of research. Even when information is not readily available, community forums and social networks enable people to access expert help and opinions within minutes.

Make magazine and its website popularized the trend in the mid '00s and have grown from 22,000 subscribers in 2005 to more than 100,000 currently. In the process, the founders met so many interesting folks that they now sponsor annual Maker Faires in San Mateo, Detroit, Austin, Kansas City, and New York. Each faire attracts thousands of people, eager to discover contraptions such as the life-sized mousetrap (a Rube Goldberg contraption inspired by the namesake board game), or human-powered amusement rides.   It's a celebration of creativity and of what people can do with time, tools, and technology.

"At a Maker Faire, the attendee can explore science, technology, art, craft, and the creativity that binds all these genres together under one roof. What other event can provide knitting workshops, Lego-built cities, clothing reconstruction, and dissecting mechanical frogs all in the same day?" says Jamie Marie Chan.[77]

Many other events have opened up around the country. The Bizarre Bazaar can be found in L.A., Cleveland, San Francisco, and Boston. Renegade hosts an annual event in Brooklyn, while the Urban Craft Uprising is in Seattle.

For those who prefer to craft on their own, an extensive collection of "how-to" videos, articles, and community boards are available on craftster.org. A thriving online social network can be found on Mycraft.com. Knitters and crocheters may want to get involved online at Ravelry.com. There is even guidance on crocheting projects that should never be undertaken at whatnottocrochet.wordpress.com.

There is a spirit of sharing for many in the crafting community, both in real life and on the Internet. Faythe Levine writes in *Handmade Nation* that the crafting community is "is the most supportive group of people that I have ever encountered. Makers share their resources, do collective advertising, promote one another's work on their websites, do trades, and share skills."

Even the commercial sites are generous with help and information. One of the more popular is Adafruit.com. Founder Limor Fried makes a point to give out all plans and instructions for free. The catch? Supplies can be conveniently purchased on the site.

When it's time to sell your latest creation, many crafters have found a market through online site Etsy. Built in 2006 as a sort of homegrown alternative to eBay, Etsy is now the dominant online marketplace for handmade goods. In addition to providing online stores where crafters can sell their goods, Etsy produces videos for its website and runs workshops with the goal of turning more people into crafters, thereby increasing its user base. Sales have grown 70% over the past year and the site is getting over a billion page views *per month.* Newer competitors, such as ArtFire, Cargoh, and Zibbert, are also getting into the game.

Hippies, Hackers, and Punks

It's difficult to generalize the DIY movement. On one end, there are crafters who focus on clothing and items for home design. In the middle, there are the "everymen" who want to do weekend home-improvement and have the satisfaction of doing it themselves. At the far end of the spectrum are the makers and geniuses who invent and create their own machines from off-the the shelf parts.

Similarly, it is difficult to pin down the exact date or origin of the current wave of crafting. What we do know is that it has hit high gear in the past decade, quite possibly as a result of the financial downturn. Crafting movements throughout history tend to coincide with periods of economic uncertainty. During periods of higher unemployment, people have less money and more time. This makes DIY an economically viable hobby and sometimes a necessity. People who are genuinely talented can even turn their skills into a supplemental source of income, providing additional stability and economic resilience.

Since my father's generation, several decades of prosperity have shifted the DIY movement from need to preference. Given the right mix of conditions, it may shift back into a means of improving the basic quality of life for families. My dad made things because he had the talent, the interest, and the need. For more recent generations, DIY has been more about making a personal statement.

It is useful to understand the history of crafting to understand why the DIY culture of today is different than it was in the 50s or 60s. DIY culture as a recent phenomenon has evolved from a blend of various subcultures and counter-cultures over the past forty years. The last time crafting achieved this level of popularity was in the 70s. It was part of the "back to the earth" culture that featured the beginning of the organic food movement, latch hook rugs, macramé plant holders, and hand-dyed muumuus. This generation was among the first to make crafting a political statement. While intentions were good, overall quality and craftsmanship were considerably varied. Key influences from this era included a focus on the recycling of old materials, self-sufficiency, and the cultural values of sharing of tools and techniques, and the idea that knowledge should be free. This last principle, in particular, was readily co-opted by the first generation of Hackers.

The Hacker Manifesto, as written in 1986, outlined the purposes and ethics of hacking – to learn, to explore, and to go beyond the limitations of modern society.

From hacker culture and the digital generation comes a belief that many of the best projects are open-source. There is a sense that you don't really own something unless you can open it and see what makes it work. Even then, everyone should be able to take apart an existing product and make it better.

Among various segments of the indie DIY movement, there is also a hint of post-modern punk attitude. In addition to being a voice of protest against commercial culture, punk attracted an ardent following because anyone could do it. It never had ambitions to move beyond the garage. It didn't take training, fancy equipment, or even *talent*. Punk in the 70s and 80s was raw and heartfelt stuff – not necessarily ready for mass consumption. The important thing was that you made your own music. Crafting can be a form of rebellion, too. As Andrew Wagner writes, "Making your own clothes, your own dinnerware, your (own) art, has become a way to politely 'give the man' the middle finger, for lack of a better term."[78]

Punk knitting sounds like a contradiction of terms, but the present generation has made it possible. In *Knitting for Good,* Betsy Greer writes specifically for the community of "radical knitters" and highlights the "subversive, revolutionary, and political aspects of knitting."[79] Betsy Greer is also the creator of craftivism.com, which features anti-war cross-stitching and projects to protest sweat labor. A forthcoming book, *Yarn Bombing* by Mandy Moore and Leanne Prain, highlights random acts of knitting as a feminist alternative to traditional urban graffiti.

It's a strange new world...

Reuse, Recycle, Repurpose

The amount of innovation that comes out of the do-it-yourself (DIY) movement is staggering. People are making things that are quirky, fun, and simply not available commercially. If you are looking to make a banjo from a cigar box, convert bacon from your refrigerator into a useable bar of soap, or turn junk mail into craft paper, you can find detailed instructions on the Internet.

There is a sense of humor found in many maker projects that sometimes borders on the absurd. At the same time, however, it highlights our ability to adapt to change and find creative solutions using an unexpected range of materials. It goes beyond reusing things or melting them down and recycling. Objects are repurposed in entirely new ways that inspire the imagination.

Jordan Bandersnatch is a crafter who specializes in recycling found materials into fashionable accessories. She makes men's wallets and dainty-yet-tough "combat purses" using bicycle inner tubes. Using scrap pieces of rubber, she makes feather earrings. She can repurpose a vintage flower print pillowcase into a child's dress, or hand-paint a collection of bottle caps into miniature refrigerator art.

Artist Brian Marshall creates whimsical sculptures for his Adapt-A-Bot Robot Orphanage. One of his "found" robots is made of old twine, a vintage toy bank, brass lamp parts, and Panasonic radio from the 1970s. On one hand, it is all junk. On the other hand, it is all art. By adopting a bot, you not only get a cute critter, you can reclaim what would otherwise be considered scrap metal.

The bestselling DIY kit at Adafruit is the Minty Boost – an iPhone battery charger cleverly cased in an Altoids mint tin.

Given this perspective of the world, virtually anything can have use, value, and aesthetic merit. It is a type of creative alchemy such that once-throwaway things are transformed into something beautiful.

Kitchen scraps are fed to a compost pile, which then nourishes a rose bush. A pair of broken mechanical watches become cufflinks. Coke cans are melted down and turned into amazingly durable aluminum chairs. Nothing ever ends; everything gets a second chance at life. There is something encouraging about all of this. It is a type of radical and local self-reliance. The finished products can sometimes be beautiful and sometimes very marginal, making the skill of the craftsman the most valuable material of all. Asking the question "what else could this be?" can be remarkably transformative.

Sometimes this takes creativity and other times it takes a certain type of innocence.

Here is another story from Cambodia. In the capital city of Phnom Penh, people set out all their waste glass and scrap metal at night. Each day, before dawn, the scrap peddlers would go through and clean the streets of anything that could be recycled. One morning, they saw a mass of unused wire, electronics, and clay at the base of one of the city's statues. The peddlers stripped away the valuable copper cable, but left behind the clay – defusing a bomb and unwittingly saving the city from an act of terrorism. Locals refer to this as "Cambodian Homeland Security".

Homemade and Hip

I went to an Anthropologie store with my wife Stephanie last month and was surprised to see an entire rack of frilly aprons in the apparel section. Aprons! Beautiful echoes of days gone by, reminding me of apple pie and homemade pastries. Since when did aprons become fashionable again? What kind of zeitgeist is Anthropologie tuning into?

Homemade is hip again, and a trip to the artsy Williamsburg section of Brooklyn will reveal masses of trendy twenty-somethings dressed in flannel and homemade accessories. A girl can be caught knitting at Starbucks, while her friend boasts about her collection of handbags that she sewed together from recycled felt.   Both wear tatty converse sneakers, tattooed with a magic marker. My last visit to the area felt like a time machine back to a somewhat ragged and edgy era.

So far, over 50,000 have signed the pledge at BuyHandmade.org, promising to "buy handmade for myself and my loved ones and request that others do the same of me."

Status is now about story – about where materials were found or how things are made. The story could be about the artist, or the amount of time and effort that it took to create something.   Brands may have trouble competing in this type of environment. Meanwhile, we could continue to see a resurgence in the folk arts, craft shows, and farmers markets. All of these create a level of interaction not found on the Internet or in a chain store. The conversation with the artisan is as much a part of the transaction as the actual purchase. For smart craftspeople, building consumer relationships isn't about loyalty cards and promotional deals so much as it is about creating networks of friendships.

Smart retailers will focus on craft, or perhaps enable consumers to create and modify existing products. They may recognize that many consumers are becoming empowered to make their own products and can establish a relationship with their customers by providing the means or inspiration to craft things in their own fashion.

More Emerging Trends:

Home Economics and Shop Classes – Back Again

For every social trend, there is a counter-trend. In this case, the era of digitization may give way to a trend toward hacking reality. What does this look like? It means breaking the code for brewing a better beer or modifying a pattern to make the perfect child's dress. It is about using knowledge found on the Internet and bringing it home in a very real way.

The online DIY movement seems to be catching on with Gen X and younger (the average age of sellers on Etsy is 34). It is a way of taking a limited budget, adding some personal time, and making something unique and cool.

Home economics, along with traditional shop classes, may come back in a big way if the trend toward DIY becomes a necessity. The next iteration of home economics classes could shift toward traditional domestic skills (such as bread making, canning, preserving, and gardening), with more of an emphasis on hands-on learning. Getting away from gender stereotypes and focusing on self-reliance will make these classes more appealing for guys, as well.

The Makerbots are Coming:

"We've had this merging of DIY with technology," says Bre Pettis, co-founder of Makerbot Industries, in Brooklyn. "I'm calling it Industrial Revolution 2."[80] The Makerbot Thing-O-Matic is a three-dimensional printer for the home that uses fused deposition modeling to rapidly prototype objects. In other words, it prints stuff.

Over the past decade, 3D printers have fallen in price from tens of thousands of dollars to entry-level models that now run in the thousand-dollar range. 3D printers now come in three basic flavors – printers that make things that are quick and easy, detailed, or durable.

3D printing has the potential to be a game-changing technology by moving productive capabilities from the factory to the desktop. The means of production becomes democratized and mass customization becomes reality. While it is currently used primarily for rapid prototyping and small-scale manufacturing, there is no reason to believe that 3D printing won't become a disruptive force in many industries. Already it is getting rapid acceptance in areas as diverse as dentistry (custom-made crowns and bridges), jewelry (castings and molds) and the automotive industry (remanufacturing parts that are no longer available). A company called Bespoke Innovations out of San Francisco now uses 3D printing to create prosthetic limbs to order.

When combined with advanced scanning technologies, another potentially disruptive technology can be developed – 3D copying. Imagine a world where you can download plans for just about anything over the Internet, modify it on your computer to your personal needs or specifications, and then hit the "print" key. Creative capital will trump physical capital and we will be entering into a truly post-industrial culture.

Open-source visionaries are taking this all a step further with the creation of RepRap machines – shorthand for replicating rapid prototypers. These machines may ultimately be able to print *copies of themselves*. So far, about 80% of the parts of a RepRap can be made using a parent machine (some assembly required).

Machine Shop or Social Club?

Also, look for hackerspaces to become more popular, particularly in urban areas and college towns. These are workshops where enthusiasts who share a passion for technology can meet, socialize, and collaborate. Think of it as a community garage where people can work on big projects with industrial-strength tools.

For a monthly membership fee, people obtain access to higher-end 3D printers and scanners, computerized milling machines, lathes, plasma cutters, and other heavy equipment to build their own projects. More importantly, it creates an opportunity to share ideas and get help from fellow makers.

While some events at hackerspaces are purely educational, such as workshops and presentations, other functions are purely social, including game nights and parties.

From Fantastic to Functional

The DIY movement serves as a type of cultural memory back to the times when self-sufficiency was commonplace. It represents a body of knowledge that can be passed from one generation to the next.

While the tools have changed and information has become more accessible, DIY can feel like a "step back in time" when everything that was old is now new again. It represents one of the great touchstones of American culture – the dreamers and garage mechanics who create entire new industries out of sheer imagination. This sense of innovation is critical to sustained recovery. It goes beyond simply getting things done to getting things done better than before.

The joy of invention is often its own reward. There is a sense of play at work, a willingness to experiment and make mistakes. The same skills developed creating simple games and toys can later be used to create things that are much complex or useful. Before the airplane was invented, the Wright brothers were simply kite makers on the beach.

If we see a sustained restructuring of our way of life, DIY may provide us with the skills to recreate some aspects of our current standards of living. But it can do so with a level of depth, diversity, and integrity that exists only in our collective imaginations.

Online Resources:

Craftzine: A modernized take on traditional crafts

www.craftzine.com

Craftster: Extensive collection of "how-to" videos, articles, and community boards

www.craftster.com

Etsy: Largest online marketplace for crafters and buyers.

www.etsy.com

Hackerspaces: Everything you need to build, manage, or find a hackerspace

www.hackerspaces.org

Instructables: 10,000+ directions on how to do anything, from making a camera to building furniture

www.instructables.com

Make Magazine: Quarterly magazine of do-it-yourself projects and instructions

www.makezine.com

MyCraft: Online social network for the sharing and marketing of crafts

Mycraft.com

ReadyMade Magazine: Quirky yet entertaining DIY projects for the home and kitchen

www.readymade.com

Chapter 5

Homesteading

Doug Fine was a journalist determined to break his oil addiction. For him, this meant packing his accumulated stuff into his car and getting off the grid. He bought a forty-acre plot in a remote part of New Mexico, which he calls The Funky Butte Ranch. It took over a year to complete the transition, learning the skills to live as locally and self-sufficiently as possible. With no farming, electrical, or mechanical skills, Doug was "green" in every sense of the word.

His goals were ambitious: to be largely self-sufficient for food, energy, and transportation. He wanted to buy locally anything that he could not produce himself.

Within the first three months, Doug converted his diesel Ford F-250 to a veg-oil rig, size XXXL. His truck has two tanks for fuel – one for conventional diesel and the second for filtered cooking oil, which he recovers from a local diner. This enables him to travel up to 1,300 miles between fill-ups. While this makes it possible to drive for free using recycled fuel, Doug mentions that "there is a downside to all of this, of course, which is if you are going to switch on the vegetable oil – then don't drive hungry, because the exhaust smells like kung-pao chicken."[81]

Doug wanted to keep his creature comforts, the most important of which was ice cream, which he considers to be its own basic food group. Going through Craigslist, he found a pair of Nubian goats, a breed known for producing milk with high fat content. His Funky Butte ice cream now provides dessert for his family, with main courses provided by his ranch's eggs, produce, cheese, and yogurt.

Becoming self-sufficient involved climbing a somewhat hazardous learning curve. While learning how to protect his chickens from coyotes, he got a case of "scope eye" from the repercussion of his rifle – a scar that took months to heal. Assembling a passive solar hot-water heater resulted in second-degree burns. He felt that adding the roof panels for power generation was best left to a professional contractor.

"But for all the mishaps, I have reduced my electric bill by 80% and no longer need gas stations to drive. All while keeping my Netflix, my Internet, my fridge, washing machine, and most of all, my booming subwoofers."

As he writes in his book, *Farewell My Subaru,* living a simple life also has its benefits. "I was happy. Happier than I can ever remember being in my adult life. I had rediscovered a childlike joy in the very real-life endeavor of living green."

Every day, more people like Doug are opting out of the system and becoming self-sufficient. Just as importantly, we are seeing a slight perceptual shift away from the household as a unit of consumption and greater emphasis on the household as a unit of production. In a very traditional sense, this is what homesteading is all about. It is a shift toward radical self-sufficiency. In some ways, Doug Fine is living a new/old American dream.

There is a growing movement among many in the U.S. to move back to the land – to unplug and a live a real life away from utility bills, commutes, and the stresses of cities and suburbs.

In many ways, the homesteading movement is grounded in a need for greater security and a desire to return to more basic times. A century ago, roughly half the U.S. population was engaged in farming. Today, it is less than 2%. Many of us might not have direct experience of farm life but have parents or grandparents who do. We still grow up with barnyard stories about animals and weather.

There are four reasons why we'll see more people "growing their own".

Safety: There is a growing movement toward organic and locally produced foods. Concerns over pesticides, fertilizers, and factory farming have all contributed to disillusionment with commercially grown foods. People are getting tired of food with questionable labels or ethics. There is a trend toward "eating clean" and consuming foods that are unprocessed and without preservatives. A basic rule of the clean food movement is that if you can't pronounce or spell an ingredient, you probably shouldn't be eating it.

Meanwhile, U.S. sales of organic foods have grown from $1 billion in 1990 to $26.7 billion in 2010 – a compound annual growth rate of close to 18%.[82]

Nutrition: Food designed for transport and refrigeration looks and tastes different than food that is bred to be tasty and healthy. Just because a tomato can resist a home-run hit with a baseball bat doesn't mean that it is the best choice for a salad.

Spinach is a good example of why locally grown food can be better for you. Spinach refrigerates well – it can be kept from going bad for up to three weeks after harvest. But after the first week, it loses about half of its B vitamin content.

Money: Nothing can be more local or affordable than growing food in your own back yard. The average American household spends $6,000 per year on food, not including the transportation costs of driving to the grocery store.

Homesteading

In a world of rising food inflation, sometimes it is cheaper and more fulfilling to simply *grow your own*.   Over a ten-year period ending in 2010, corn prices have gained 180%, rice 135%, soybeans 179%, and wheat 184%.   This means that prices for three of these basic food commodities have almost tripled over the past decade. Growing your own can contribute significantly toward a sense of personal stability.

There is another advantage – homegrown food is less likely to spoil within days of harvesting. Food growing on the vine doesn't need to be refrigerated and is fresher than you'll ever find in the grocery store or farmer's market.

Time: With an official unemployment rate of close to 10% and the leading edge of the baby boom now hitting the traditional age of retirement, there will be more free hands with time to dig in the dirt. A slow economy is good for productive hobbies and interests of all sorts. An increase in home-based businesses and the added convenience of telecommuting means that more people may be found at home, tending to their vegetable patches.

Preference: Nothing tastes better or fresher than something that you've grown on your own soil. Growing the juiciest tomatoes or the biggest pumpkin has been the source of bragging rights for generations.

There is simply no comparison between a homemade cherry pie and one purchased from the grocery store and made with canned and dyed filling. Nothing says, "Welcome" to a new neighbor better than fresh baked cookies or a bouquet of homegrown flowers.

Home-brewers have known for decades that if you want something special and unique, sometimes you need to do it yourself. Last year, my brother-in-law gave family members samples of his best home-brewed molasses whisky, flavored with anise – giving it a slight taste of licorice. These were poured into unlabeled blue bottles, corked, and sealed with melted wax. The potency of the concoction (185 proof) made it the conversation and toast of several gatherings.

In the emerging culture, social status isn't necessarily about what you have, but what you can make or do. It may be far more relevant and helpful to know someone who can harvest his or her own honey than someone who knows of the best resort in a faraway country.   In the event that we see a further spike in food prices, you are going to want to have good relationships with your neighbors -- to share what is in abundance and find or make what is no longer available in the stores.

If we see this change in value systems, it could help vitalize a sense of respect for our elders. They have personal experience with many of the skills that younger generations are attempting to rediscover.   They are often willing and quite able to offer a helping hand for many of these homesteads.

What interests and fascinates us seems to be shifting away slowly from high-tech to low-tech.   As we make this transition, conversation may shift from the abstract to the very practical. While we still will have access to the Internet, we might use it more as a tool to give us knowledge for basic living. It will always provide a source of news and entertainment – that part will never go away. In addition, it will also provide a connection for others to share what they are learning and doing.

The transition toward homesteading and "growing your own" is a natural counter-trend against a decades-long shift toward an increasingly digitized and consumer-driven economy. There is a certain satisfaction in pointing to a rack of newly bottled wine or a row of freshly canned preserves and saying, "We did *that*." It feels much more real than manipulating bits and bytes on a computer, or preparing a complicated tax return.

The things that we aspire to do and have will change also. Conspicuous consumption may increasingly appear insensitive and inappropriate.   In the emerging culture, creative solutions to everyday problems will win more friends and approval.   The novelty of having the latest car or gadget just isn't as important when you are trying to put food on the table. People simply might not be able to afford the latest, flattest TV set, but they can learn how to bake a sweet potato pie.

As part of this transition, the things that we may look for in a home will evolve with the times. Large kitchens may still be practical (and more frequently used), but overall floor plans may become more compact. Tool sheds may replace pool sheds. Garages may once again be used to store machinery and equipment.

Overall, the home becomes more purposeful and more productive. New homebuyers might ask questions like "Is there room for a home office where I can meet with clients?" Or, they might ask for landscape elements such as fruit-bearing trees. Local sources of energy generation, including solar panels, geothermal wells, or wind turbines, might also become an important consideration when buying an existing home. Greenhouses would appeal to people who are looking for a year-round supply of fresh produce, while cisterns and rain barrels could help offset water needs.

Houses cost money. Homesteads create independence.

Microfarms are Big

Front-yard farming is bound to create some controversy among established civic associations, but may become commonplace if there is a continuation of food inflation and/or shortages. Just consider that during World War II, twenty million Americans had "victory gardens" in their backyard. These grew 40% of our country's vegetable supply.

Burpee Seeds was a rare company that saw booming sales during the first financial crisis, reporting an increase in sales of 60% between 2007 and 2008.

Rachel Hoff and Tom Ferguson started their garden five years ago and the project kept getting bigger. "We'd add more beds and it was never enough," Rachel says. When the economy stumbled, real estate had become more affordable. "At that point, we had chickens and we realized we could do more. We were looking for a small house on a big lot." What they found was a 750 square foot home with a quarter-acre lot in the city of Vallejo, California. They named it Dog Island Farm.

With the housing bust, however, came a reduction in income from their construction jobs. So, in 2010 they began an experiment to live a year without taking a single trip to a grocery store, convenience market, box store, or restaurant. They would allow themselves to join a community-supported agriculture program (CSA) and make trips to the local farmer's market. They could also eat what they grew and barter for what they couldn't produce themselves.

Rachel and Tom chronicle their story on a blog site called *A Year Without Groceries.*

According to Tom, "Our whole thing in doing this is not to go broke." They estimate that they saved $6,500 in food and spent $2,500 in improvements last year. Despite an unusually wet season that destroyed much of their backyard fruit crop, Dog Island produced 1,800 pounds of food, including meat from turkey, goat, rabbit, and duck. They also raise bees for honey, and grow a mind-boggling array of heirloom vegetables.

Tom and Rachel like knowing where their food comes from. Whether they are consuming their own livestock or trading with other farmers, the freshness and quality is higher and they feel a closer connection with their food. As Rachel says, "You can meet the animal and see how it is being treated. A lot of people have taken the faces off of their food. It comes from a living creature. We know how they have been fed, we know how they have been treated, and we know how they have died. "

"We have all the chicken and rabbit we could ever want. We make all our compost now. Our kitchen scraps go right out into the chicken yard and back into the garden. The chickens are the composting kings and queens. There is no way to produce enough compost without them. They act as pest control and they make a huge contribution to soil."

While they have a productive farm, they don't have the necessary licensing or permits to sell their food. So they live off their own farm and preserve as much food as possible through canning, drying, and pickling.

"Our neighbors have been very supportive. We give them anything extra that we have, which they really appreciate. Communication is so important when you are doing this with neighbors right next to you."

Home Schooled and Happy

Tight school budgets and concerns about the quality of public education are encouraging many families to become responsible for their children's learning, as well. In the span of eight years, the number of children educated at home has increased by 75%.[83]

This is still a niche trend and motivations for many families vary. Some desire to teach moral values to their children, while others are more focused on providing a healthier learning environment and immersive curriculums.

After medical school and residency training, Tom and Christian Wathen decided to leave the congestion and stress of the East Coast for a quieter life in Mathews County, Virginia. It is a rural area with more veterinarians than doctors. The land was comparatively inexpensive, which enabled them to trade in their subdivision home for a 10-acre mini-farm. While Christian supports the family as a physician, Tom manages the property and teaches their teenage twin daughters. They also have a younger son who elected to attend public school with his friends.

A typical day for the Wathen kids starts at 6:30 to feed and water the animals. They start home schooling at 8:00 and continue through 12:00. A half hour each day is dedicated to biblical studies. There is an hour after lunch for music practice, and another two hours is spent doing chores around the house (mucking stalls, weeding the garden, cutting/stacking wood, cleaning the house, etc.)

The girls have some free time in the afternoon and then play sports in the early evening. Between activities at the YMCA, 4H, Little League softball, church youth events, and field trips with other home-schooled children, there are no shortages of opportunities to interact with other teenagers.

Tom submits his curriculum to the state each year, and his children are required to take the Virginia Standards of Learning tests to monitor their progress. The Wathens typically score within the top third of their peer group.

Melissa and Kathryn are beautiful kids who feel that they have the best parents in the world. Unlike many teenagers, they seem to radiate a sense of peaceful calm. They easily participate in adult conversations and are unusually polite. When it comes to family, Tom says, "You get what you put into it."

If time is available for the parents, a solid case can be made in favor of home schooling from a learning standpoint. In a landmark 1998 study by Lawrence Rudner at the University of Maryland, the standardized test scores for over twenty-thousand home schooled students were reviewed and compared. The median test scores of home-educated kids of all grade levels were higher across all grade levels relative to public and privately schooled children. What makes this even more interesting is that over one-quarter of the home schooled students were enrolled in a grade level higher than they would typically be based on age alone.

On average, homeschooled students at elementary school levels perform one grade level above children their age on achievement tests. This gap tends to widen after the fifth grade.[84]

Home schooling is comparatively inexpensive, too. According to the study, the median amount of money spent on homeschooled education was $400 per student. Home schooling doesn't take much money -- it takes time, interest, and lots of patience. Public education isn't going away anytime soon, but if our country continues to experience elevated levels of unemployment combined with accelerating inflation, we may see more families consider this alternative for their children's education.

Staying Connected

For all of the charms of the rural lifestyle, people who move into homesteads quickly realize the importance of having good relationships with neighbors. Not only are supplies and Wal-Mart far away, but very often there are big projects that need to be done. Many hands make for light work.

It is odd to reflect that urban living can create a sense of anonymity, while country life can sometimes promote a higher level of communication – often out of necessity. In comparing the two lifestyles, there are clearly pros and cons to any given choice from a sustainability perspective.

Unless we have a complete economic meltdown, however, it may be difficult for rural homesteaders to economically stay ahead of their counterparts who stay closer to urban areas. The opportunities for income will remain higher where there is high population density. Perhaps just as importantly, cities and suburbs allow people to *specialize* and perform work that best fits their skill set. Years of training, education, and experience simply make people more productive at certain tasks. There is a huge economic value to this. If my wife and I were to live off the grid to become goat herders, our overall economic productivity would be significantly lower than it would be if we kept our jobs as professionals.

Just as importantly, interactions with larger numbers of people create seemingly random encounters with other people who are also very good at what they do. It's long been known that there is a broader range of opportunities in urban areas, where it is easier to stay in touch with former business contacts and meet new people.

One homesteader, Jennifer Dukes Lee, shares in the High Calling blog that:

> *We moved to the farm in 2002. Even though I was four hours from the metro newsroom, my editors kept me on staff. But I quit a couple years later. I felt that my writing and reporting were suffering because I was so far removed from the daily news operation and the camaraderie of reporters. So, this became a wilderness time for me.*
>
> *But it wasn't all bad.*
>
> *In my wilderness time, I roamed about. I sought joy in other things, simpler things. It was a time of intense soul-searching. I suppose that some in the news business would still say I'm roaming aimlessly about in the wilderness. But I rather like it out here, for I'm finding the wilderness has a joy all its own.*[85]

For all of its simple pleasures, going completely off-grid may have hidden costs and drawbacks. This is why the Wathens and the folks at Dog Island Farm seem to have the best of both worlds. Both have a degree of independence and self-resilience, but they've kept their day jobs, too. As a result, their lifestyles offer a greater range of choice and economic stability.

More Emerging Trends:

Edible Landscaping

Landscapes can be beautiful *and* productive. As people contemplate the uses of their property, more are starting to consider dual-purpose plantings. Instead of a boxwood hedge, people may choose to plant raspberries, blueberries, and blackberries along their fence lines. Container gardens filled with kitchen herbs are also coming back in popularity.

Fruit trees can provide shade, good looks, and food year after year. Some varieties are available on a dwarf rootstock. While the fruit is a normal size, the root limits overall growth to a manageable height and makes for simpler picking and maintenance.

"Edible landscaping has just taken off, particularly as people are concerned about food security," says Katie Fraser of Raintree Nursery in Morton, Washington. She also mentions that the slow food movement has opened people to the many varieties of fruit that they can grow. While there might only be a half-dozen types of apple at the grocery store, there are hundreds of different varieties that can grow in the U.S.

To get the best results from an edible landscape, Fraser recommends that people contact their county extension office to learn more about which plantings do well in their area.

Up on the Roof

Front and back yards are not the only option for new gardens. The best location might be right over our heads. In New York City alone, there are now at least five different rooftop farms.

Gotham Greens recently opened a 15,000 square foot facility in Brooklyn that will produce year-round crops under a greenhouse roof. They are expected to produce 100 tons of vegetables per year, including arugula, bok choy, swiss chard, and three varieties of lettuce. The opportunities for expansion are significant. "There are literally thousands of acres of unused rooftop space," says Gotham Greens co-founder Viraj Puri.

The Brooklyn Grange operates a 40,000 square foot farm on a factory rooftop in Queens. They use a root-barrier to protect the roof, drainage mats and plant cups to handle overflow from rain, and a soil blended from compost and lightweight, porous stones.

Rooftop farms are new frontier for agriculture, and there continues to be significant experimentation in this area. Container gardening is a relatively inexpensive and quick option for restaurants and apartment dwellers. Meanwhile, the bigger greenhouses tend to focus more on hydroponic methods, which deliver nutrients to plants without the use of soil. Water is recycled constantly, and there is no soil erosion or run-off. Hydroponic greenhouse growing reduces the risk of infestation and disease and also allows for control of temperature, humidity, and lighting conditions. Companies such as Valcent are pioneering the use of growing racks to make the hydroponic process even more efficient and productive.

Dickson Despommier from Columbia University sees urban farming going vertical and has proposed multi-story projects for skyscraper farms. "It makes sense to grow food where people live," he said on CNN. In a vertical farm, there are multiple sources of revenue. There might be a retail grocer on the ground floor, a penthouse level restaurant, and ticket sales from agri-tourism.

Foraging Foodies

Field trips merge with grocery shopping as more people look to the wild for edible herbs and mushrooms. This trend is still off the radar for most, but cutting-edge chefs such as Matt Lightner at Castagna Restaurant in Portland (Oregon) and Daniel Patteron at Coi in San

Francisco were among the first to put wildcrafted greens into the frying pan.

Other restaurants, such as Eleven Madison Park in New York City and Alinea in Chicago, are now serving wild ingredients such as toothwort, cornelian cherries, sweet cicely, and licorice fern.

"It's the frontier," Lightner says. "The woods are this mysterious area where things grow. You don't have to tend it, and you don't have to plant it, you just have to find it. Everybody is used to exotic products you ship in, or the farm-to-table thing. Now people have an interest when we serve them something they spotted when they were out on a hike."[86]

The Electronic Schoolhouse

Distance learning is quickly becoming mainstream, and some public school districts are learning how to embrace the trend. While the charter school movement has been growing for the past few years, cyber-charter schools may now be the leading edge.

Cyber-charter schools use a state-approved curriculum to deliver education over the Internet to students at the time and location of their choosing. It is a trend from which students, teachers, and schools can benefit.

Instead of spending time in classroom management and discipline, teachers are now working with students on a one-to-one basis. In the new system, students send an e-mail instead of raising their hand. Teachers are required to respond within 24 hours.

Overall class sizes remain the same. School districts save money by hiring non-union teachers and by avoiding costs of maintaining physical school buildings and a bus system for students. Currently, twenty-seven states allow full-time online learning. In Pennsylvania, the costs of maintaining a cyber-charter school are 30-40% lower per student than for a traditional brick-and-mortar system.

Parents still play an important role in supervising and monitoring their children from home. The Pennsylvania Leadership Charter School is experimenting with "electronic schoolhouses." In this format, a parent is paid by the school district to help facilitate child education for neighborhood children from a nearby facility, such as the local church. This creates a mixed-aged classroom that gives older children the responsibility for mentoring their younger peers.

Online Resources:

Backwoods Home: Self-reliant living and preparedness for families and homesteaders.

www.backwoodshome.com

Enchanted Learning: Subscription-based curriculum for homeschoolers provides over 30,000 pages of content for languages, science, history, and art.

www.enchantedlearning.com

Keystone School: Accredited online home school education program.

http://keystoneschoolonline.com/

Hobby Farms: For those who are making a living or a life from their farm.

www.hobbyfarms.com

Books:

Mini Farming Self-Sufficiency on ¼ Acre by Brett Markham

Urban Homesteading: Heirloom Skills for Sustainable Living by Rachel Kaplan and Ruby Blume

Surviving the Apocalypse in the Suburbs by Wendy Brown

Real Goods Solar Living Sourcebook: The Complete Guide to Renewable Energy Technologies & Sustainable Living by John Schaeffer

Chapter 6

Community

The village of Arden is one of the oldest surviving utopian communities in the U.S. It is easy to miss, because the residents prefer to live there without announcing themselves to the world.

Much of the area is covered by trees. Few of the roads are well paved and there is almost nothing in the way of outdoor lighting. It gets dark at night, but you can see the stars. Most of the homes are modest cottages, but artistic and often handcrafted.

Ardenites are creative and self-amusing. The heart of the village is the Gild Hall, which hosts weekly community dinners, regular dances, and live music events. Just outside is an open amphitheater for the performance of Shakespeare plays by resident actors. Across the street is the Buzzware Village Center, where the Georgist Gild meets to discuss utopian economics and the Gardening Gild maintains their flowers and vegetables.

There are many shared open spaces in Arden, including a village green and Sherwood Forest. The town maintains a single tax based upon the square footage of one's property lease. This means that most homes occupy small lots, clustered close together.

Arden was founded in 1900 as an experimental community based on the ideas of Henry George and the values of the Arts and Crafts movement.   It grew so popular that two adjacent communities, Ardentown and Ardencroft, were formed under the same guiding principles.

Today, Arden exists as kind of an altered state. While visiting, it is easy to feel relaxed and lose track of time because the pace is so different. When a local reporter recently asked if the forest is inhabited by fairies, several residents responded, "Of course it is" and then wondered why anyone would even bother to ask.

Many Ardenites are now aging Boomers.   Fun, open-minded, yet opinionated, they engage easily in conversation and socialize well.   The night after Halloween, the adults go door-to-door, as neighbors fill their empty beer steins with homebrew.

The village is resilient. People live close to one another and are better for it. They work together on numerous events and projects throughout the year, including the annual Arden Fair and a recent renovation of the one-hundred year old Gild Hall.   The community is self-sufficient in providing access to art and entertainment. While they have their own community garden and library, they still like to shop for groceries at Trader Joe's and buy books at a nearby Barnes and Noble.

The "real world" is not so very far away. If you drive a mile or two out from the village, you will see the strip malls and fast food chains that dot the American landscape.   Drive a few more miles, and you will see the same corporate restaurants and "big box stores" appearing once again, in remarkably similar shopping centers.

The Arden Gild Hall is what urban ethnographers would refer to as a "third place." While most people spend the bulk of their day at home or at work, the time spent in public places is mostly limited to shopping, church, and Starbucks.   In his book *The Great Good Place*, Ray Oldenberg emphasizes the need for third places as important for democracy, civic engagement, and cultivating a sense of identity.

Third places are the anchor of community life. They allow for interactions with other people, and opportunities for building relationships. Oldenburg writes that the best third places include inexpensive (or free) food and drink, a high degree of accessibility (preferably within walking distance), and "regulars" – people who show up consistently.

In the old days, general stores, post offices, and local pubs served as secular third places. More recently, coffee shops have become popular places for people to congregate and interact. However, shops that provide Wi-Fi Internet access may create an empty sense of togetherness, as patrons are less likely to interact with each other. Chances are, they are checking in on their Facebook updates or their Twitter feed – not chatting with the person sitting next to them. In a crowded coffee shop, you will find many people who are alone together.

Go to an airport and watch as everyone seems distracted by their iPhones, Androids, and Blackberries.

At the mall, teenagers seem to be talking to each other less and less. They sit together quietly and text messages to each other, periodically giggling to themselves as they have their own private conversations.

A year ago, my wife and I went to a nice restaurant on the riverfront. A family sat down at a table next to ours – two geeky parents and a pair of nerdling children. All four of them simultaneously popped open their laptops as they waited for their menus. They seemed perfectly happy and comfortable tuning each other out.

The biggest problem with being online in the digital age is the risk of becoming *disconnected* from the real world. Attention deficit disorder, autism, and constant partial attention are all now a part of the modern lexicon. We are all still *here*, just not as *present* as we used to be.

For every trend, there is a countertrend. The countertrend to the digitization of everything is a return to the tangible and the real. It is a reoccurring theme of this book: People are seeking balance in their lives. During times of discomfort and uncertainty, people just *feel* better when they can take a certain amount of responsibility for themselves. This theme shows up in the trends toward homesteading, DIY, new emphasis on the family, and local economics. It explains why people prefer local microbrews and artisanal breads, and why small farms are booming. The macro-trend is about balancing the global and the digital with the local and the physical.

This extends into building strong local networks and friendships. Here are a few more reasons why it makes sense getting to know your neighbors:

Social Life: While everyone is busy and arranging formal get-togethers can be difficult, it is sometimes nice just to chat for a moment while walking the dog or going to the mailbox.

Pooled Resources: Having good relationships with neighbors creates more opportunities to share. We've had some of our best meals and conversations at our monthly neighborhood potlucks. People come for the good company and because it is easier to make one great dish than to cook several different courses. It is cheaper than going out to a nice restaurant and a whole lot more fun!

Safety: It is comforting to feel like you can depend on your neighbor in the event of an emergency. People who participate in their neighborhoods put more eyes on the street, making it less welcoming to potential burglars and vandals.

Beauty: People who cultivate good relationships with their neighbors are going to be more likely to spend time outside and are more concerned with the appearance of their home and property. If people like where they live, they are going to spend more time keeping their property attractive and welcoming. This creates a sense of community pride.

Strong communities connect people with opportunities and support. Perhaps more importantly, having good relationships with immediate neighbors can make life pleasant and enjoyable.

Times of change and turmoil can bring people together and build stronger communities. When people look back at the "good old days," they are usually reflecting back upon the times they felt connected to their families and neighbors. We may reminisce during the coming years and realize that these, too, were the good old days.

More Emerging Trends:

While we may see an increase in neighboring, there also exist several growing movements that indicate a growing demand for greater community engagement.

Slow Towns

"A good city is like a good party – people stay much longer than really necessary because they are enjoying themselves." – *Jan Gehl, architect*

The Slow Town movement started in Italy as a way to preserve the best of European culture. In slow towns, food is produced locally, cars are banned from some streets, and outdoor living is encouraged through ample lighting, outdoor cafes, and abundant plantings. Walking and biking are the dominant forms of transportation.

Don't bother looking for a McDonald's in a slow town – most restaurants are locally owned and often showcase regional specialties.

Co-Housing

The concept of co-housing started several decades ago in Denmark but has recently gained some popularity in the U.S. Currently, there are one hundred fully operational co-housing communities in the U.S., with about as many in the planning and building stages.

Co-housing is a model of collaborative development "in which residents actively participate in the design and operation of their own neighborhoods."[87] Most co-housing communities are designed so that houses face inward toward a shared common green, with streets and garages on the outside periphery. This creates a more walkable neighborhood with greater opportunities for social interaction.

Small, private dwellings have all the elements of conventional homes, but co-housing communities also have a larger shared "common house." The common house contains shared resources such as a community dining room, a workshop, a media room, or a library. Other shared resources may include a playground or community garden.

For most co-housing communities, group dinners are a regular excuse to share good food and enjoy the company of neighbors.

Transition Towns

The Transition Town movement was started in 2004 in Ireland by permaculture expert Rob Hopkins. The Transition, in this particular case, refers to a shift toward self-supporting communities. Hopkins felt that while individual efforts to create resilience were too small, most government programs were too slow. The best response may be at the community level.

Rather than attempting a single answer to the problem of making communities more sustainable, Hopkins developed a process by which community members could envision and then work together toward creating a more sustainable future.

While co-housing tends to focus on new housing construction, Transition Towns almost always work with existing homes and infrastructures.

Each group is a grassroots effort, based on the needs and resources of each community. Participants are encouraged to form alliances with other groups and their local governments.

Transition Towns tend to use many of the strategies outlined in this book. Members play an important role in developing local for community-supported agriculture, local currencies, seed swaps, urban orchards, and reskilling classes.

There are now over 400 Transition Town initiatives worldwide.

Online Resources:

How to Be a Good Neighbor:

www.wikihow.com/Be-a-Good-Neighbour

www.ehow.com/how_3428_be-neighbor.html

Fellowship for Intentional Community: A useful directory of Intentional Communities

www.directory.ic.org/iclist/

The Cohousing Association of the United States

www.cohousing.org/

Miiu: Resilient community wiki

www.miiu.org

Transition United States: Find or start a local Transition Town

www.Transition.us.org

Chapter 7

Sharing More, Shopping Less

WELL-TRAINED FURNITURE SEEKS NEW HOME. Sofa, loveseat, chair, and ottoman come from same litter. They love people and understand basic verbal commands such as "catch", "no", "stay", and "play dead." These loveable critters have been housetrained for 8 years and are slightly frayed -- but without tears or stains. Immediate pick-up (by Tuesday) needed. Free to caring home.

After leaving my last job, it was apparent that I would be spending more time writing at home.   So we decided to get some new furniture and soon realized that we had to get rid of the old stuff first.

We didn't want to throw out the furniture. It was very useable and looked perfectly fine from a moderate distance. Furthermore, the local landfill was already the second highest point in the state (we live in Delaware) -- we had no desire to help establish a new record.

Finally, moving it all out of our 100-year old home would take some serious lifting power. Steph and I agreed, "Wouldn't it be great if someone would just come in take this out?" That morning, we put a post online at Craigslist and received a call just five minutes later. The call was from a guy in his twenties who had just purchased his first home and had no money left over. Within two hours we received four more inquiries.

We also signed up on Freecycle to move a few other things. Ever since, we've been getting daily updates on what is available in the area. This week, there were offers for a Keurig Coffee maker, a stereo system, a 10-speed bicycle, and a Toro Lawnmower.

It felt like we had become a part of a secret club that was dedicated to giving away free stuff to members. How cool is that?

The "secret club" on Freecycle now has close to eight million members across a network of close to five thousand local groups. It is a grassroots, nonprofit movement that is committed to reuse and keeping good stuff out of landfills. Membership, of course, is free.

In the old days, sharing was done primarily amongst family. Spouses would share everything, and children would pass down their clothing and (sometimes) share their toys. For projects, neighbors might share power tools and even volunteer some time and expertise in exchange for a beer.

Sharing has moved beyond the neighborhood and into the world. For example, Neighborgoods allows users to develop sharing networks among friends, church groups, students, co-workers, apartment dwellers and anyone else within a convenient distance. Inventories of available goods are posted and searchable. They also help people keep track of everything with calendars and automated reminders.

"Business has spent centuries making buying really easy. We're just at the beginning of making sharing easy," says Neal Garenflo, editor of the online magazine *Shareable*.[88]

The boom in peer-to-peer sharing sites has created enormous opportunities in expanding both the availability and range of sharing opportunities. In the old version of sharing, things were lent only to trusted friends or relatives. In the digital age, trust is established through an online community and sharing can be done with complete strangers.

Sharing comes in two different flavors – gifting and lending. Gifting is in some ways easier than lending. Gifting is often a charitable act to help out someone in need. Sometimes gifting is done to manage extra stuff, such as food or clothing. When gifting things, there are no expectations of reciprocation or follow-through. Lending, however, requires a sense of trust. There is typically an expectation that something will be returned at some point in the future. The higher the value, the more trust is required. You would lend your neighbor's kid a favorite book, but maybe not the keys to your car.

Rachel Botsman, author of *What's Mine is Yours*, notes in her TED talk, "Technology is now enabling trust between strangers. We are sharing and collaborating in ways that are more hip than hippy."

Early in its evolution, eBay had to deal with the issue of creating trust between participants in its online auction market. Not only were customers buying from people whom they couldn't see, but often their online identities were under user names that were clearly made-up. So, eBay encouraged users to rate each other and then proceeded to post detailed transaction histories, ratings, feedback and comments on both sellers *and* buyers. If a user was slow to mail a check, sold damaged goods, or misrepresented a product online, his or her reputation would be adjusted accordingly. Developing a good reputation became an important key to becoming a successful user and completing subsequent transactions.

Online, as in real life, reputation is everything. People with a good reputation get more opportunities and develop stronger networks. Trust often starts with small transactions, such as exchanging e-mails or personal information, and eventually leads to larger transactions. Trust can also be built on the basis of referrals or ratings generated by other members of a community.

According to a study done by Latitude, online sharing often precedes sharing in the real world. Every study participant who shared information or media online also shared various items offline. The Millennial generation has been sharing online since childhood, making this group somewhat more likely to share than older generations.[89]

Rachel Botsman takes a slightly different perspective. She notes that "a lot of these are very old market behaviors, but they're being reinvented on a scale and in ways that we've never seen before.... Millennials are the foot soldiers driving this change because it's very intuitive to them; they've never known any differently. But I think that it also seems natural to the older (war-time) generations – because they had a very strong sense of community."[90]

What we are finding is that virtually *anything* can be shared. While Craigslist and Freecycle are still the best known of the online peer-to-peer sharing networks, we are seeing a baffling range of new sharing opportunities.

SharedEarth is a project that is focused on connecting landowners with gardeners and farmers. It could almost be considered as a sort of dating site for foodies or anyone who enjoys fresh homegrown produce. What people plant, the sharing of costs and labor, and how produce is divided is mutually discussed and decided.

In an interview with Treehugger, founder Adam Dell talked about why he started SharedEarth:

> *I wanted a garden, but I don't have the time or know-how to garden myself. So I put an ad on Craigslist and within a couple of days I had several responses. The ad said, "I'll provide the land, water and materials if you'll provide the work. We can share the produce 50-50." I*

found a credible person who loves gardening but lives in an apartment. We met, came up with a plan, and she got to work.

She put together some topsoil, some flower beds and seedlings. Now I have a rich, vibrant garden on my property where we grow tomatoes, jalapeños, arugula, Italian figs, and spinach. I love that I have a beautiful garden that occupies what used to just be a patch of grass and I'm getting fresh produce from it. My gardener loves gardening and loves that she gets to take home a portion of the produce.

When I started to investigate this, I called around to a number of community gardens in Austin and around the country. What I found is that virtually all of them have waiting lists. What that tells you is there are many people out there who would like to have access to land, but can't find it.[91]

SharedEarth is now the world's largest community garden, with over 5,000 acres of land shared among 52,000 participants. Malcolm Gladwell is an enthusiastic supporter, and the group has won numerous accolades.

Online sharing communities also offer extensive opportunities for venturesome yet inexpensive travel. One of the most popular is through the Couchsurfing network. Unlike the old days, when couchsurfing was only done by close family and friends, it is now becoming an international network of travelers and vagabonds looking for cultural immersion. Couchsurfing still suffers from somewhat loose connotations – but only for those who have not yet experienced the small joys of hosting or meeting new friends in faraway places.

An avid couchsurfer, Paul Murray writes, "The thing we were most concerned about finding when we were travelling was local contacts.

We didn't want to just exist within the tourist bubble - we wanted to know what actual local people were like and what it was like to live in a place."

Dinner with strangers may be one thing, but is it safe to host an overnight guest that you have previously never met? According to Paul,

> *Couchsurfing has a great safety system so that all users feel like they can make an informed decision about who they let into their homes, or stay with. Your own profile builds up verifications and vouches, depending who's met you and has decided you're trustworthy....*
>
> *What I feel about couchsurfing is that it's more than just free accommodation. It's more than just cultural exchange, even. It's good for the soul. I've found that spending time with new, generous people injects new life into me and revives my positive spirit... It shows that even with all that goes on in society today you can still rely on the kindness of strangers.*[92]

Today, there are more than 2.7 million participants in over 240 different countries registered on Couchsurfing.org.

The overall size of the sharing economy is somewhat difficult to estimate, as purchases are never made. In many cases, sharing more closely resembles recycling than consumption. While an $80 stay at a conveniently priced motel generates a modest amount of measurable economic activity, it less often generates fond memories. Couchsurfing may generate warm memories and new friendships, but relatively little in the way of taxable economic activity.

What most forms of sharing have in common is a need to redistribute surplus. This surplus can be in the form of unused property, space, or time. To the extent that relatively affluent cultures (such as the U.S.)

have more surplus, significant benefits can be generated by managing that surplus more effectively.

One way of coping with an economy in transition is to share more of what we have. This leads to a far better use of resources, less waste, and a reasonable excuse to meet and get to know people within our communities. Perhaps most importantly, it enables us opportunities to establish and re-build a collective sense of trust.

Most people think about their personal assets in terms of the size of their bank account and the value of their house. But it is useful to think about surpluses as well. If you can think of everything that you don't use every day – bicycles, camping equipment, boats, old clothing – there is a good chance that you could share some of these things without making a dent in your lifestyle.

There are also ways of monetizing unused assets in a sharing economy. Bigger-ticket items such as a car, an unused vacation home, or expensive tools and equipment can be rented through numerous online markets. We'll cover some of these opportunities in Chapter Eight.

This goes far beyond how we manage our stuff. It is about the freedom found in the shift away from the traditional concept of ownership and toward the ideal of access. It is about the amount of time saved by sharing more and shopping less. It is even about how we celebrate the seasons and the passage of time.

Culturally, we tend to binge on media-induced buying sprees each year around the holidays. In a period of economic transition, it might be unrealistic to expect that we can continue purchasing stuff at the same rate as in the past. This is not good for us, our wallets, or our planet. In a transition period, there might be a shift in gift-giving trends during the holidays. We might be more inclined to make small gifts from scratch. Or, as in Europe, we may also start to accept the idea of gifting gently used items to our friends.

One of the traditions that we keep with some of our friends is an annual "re-circulation" around the holidays. It involves putting together a gift basket of things that we've enjoyed and simply want to pass on. These could include gadgets, novelties, music, books, pictures or anything else. It can be almost completely random at times. Every year there are plenty of surprises – but a dollar is never spent by anyone.

Sharing vs. Hoarding

If you think about it, sharing is the opposite of hoarding. Writer and futurist John Renesch notes, "Fear generates the need to be protective, to hoard what one has acquired. Fear breeds scarcity and furthers the separated state of the ego. Fear leads to the need to control and mistrust anything or anyone we don't have control over."[93]

The fear-based response to times of change is to bunker down and get as much stuff as you can into your bomb shelter. That means lots and lots of freeze-dried salmon. Hoarding creates shortages. Spam flies off grocery store shelves like there is no tomorrow. Even in anticipation of winter snowstorms, milk and eggs can become scarce commodities.

Sharing often results in things *magically appearing* wherever they are needed. Anne Thomas, a blogger for *Ode* magazine, once reported about her experience in Sendai, Japan, after a tsunami. Her small home was greatly damaged, so she temporarily moved to a friend's place, sharing supplies like water, food and kerosene. There was no electricity, as the nearby power plant was destroyed. They ate by candlelight and told stories at night.

When people did have the good fortune to have running water, they would put up a sign in front of their home so that others could fill up their buckets.

As she returned to her shack each day, she would find food and water by the door, not knowing how it had arrived.[94]

Less Is More (more or less)

One of the side effects of a sharing economy is that people not only *need less*, but they can choose to *have less*, too.

Simplifying doesn't stop simply at getting rid of stuff. It extends into making new space where you put the old stuff.

Think about how much useable space in your house is committed to storage. Visualize all the filled-up spaces in your house, from the boxes in the basement, the unused but filled guest bedroom, and the heaps of unused equity and tools in the garage. Next, consider what fraction of your house is committed to keeping this stuff. Finally, look at your total housing expenses (mortgage plus property taxes plus utilities) and multiply that by the fraction of your home that is used for storage. This is how much keeping your "stuff" is costing each year!

For extra credit, you can think about your hourly wages and then estimate the cost of keeping your stuff as measured in work hours.

A lot of people seem to be running the numbers and reaching similar conclusions – stuff costs money, takes time, and consumes space. Why not just simplify?

Granted, there is a difference between choosing to downsize and being forced to downsize, but it seems that quite a few in the U.S. are going on a consumption diet. We are beginning to see some early indicators that people are starting to pare down what they feel they need to live comfortably.

The U.S. Department of Transportation noted that in 2009, the U.S. vehicle fleet shrank by 2%, the biggest decline in decades. Over 14 million cars were scrapped, while only 10 million cars were bought.[95] While part of this was the result of the "cash for clunkers" program, the total amount of miles traveled on U.S. highways has also declined since 2008.[96] Lester Brown, president of the Earth Policy Institute suggests,

"Perhaps the most fundamental social trend affecting the future of the automobiles is the declining interest in cars among young people." While car ownership was a rite of passage and key to personal freedom a few decades ago, "many of today's young people living in a more urban society learn to live without cars. They socialize on the Internet and on smartphones, not in cars."[97] The smartphone now represents communication, freedom, and opportunity.

Meanwhile, we are also seeing a slight shrinking in home size. Since the peak of the housing bubble in 2007, the size of newly constructed homes has diminished 5%, or 144 square feet (roughly the size of a child's bedroom). Many builders expect that the average size of new construction will decline another 200 square feet by 2015. The American Dream is looking a little like a Shrinky-Dink.

"Home buyers are asking for less, cutting back on options and reducing square footage," says Steve Pace of Pace Development Group. "They're saying, 'Maybe we don't need that 5,000 square footage. Maybe our bath doesn't need to be big enough for our whole family and all our neighbors to take a shower at the same time.'"[98]

The trend seems to be toward consolidating living, dining, and kitchen areas into a combined "great room". Walls are coming down and spaces are moving toward multi-purpose arrangements.

Digitize Me

Something else is happening in the American household. Electronics and appliances are disappearing without a trace. No complaints have been made. No police reports have been filed.

VCRs were replaced by DVDs, which are now fighting against unlimited video on demand. Telephones were disconnected years ago in favor of mobile phones. Fax machines are losing a battle against PDF files and e-mail. Large stereo systems are finding themselves kicked to the curb (nobody mixes music tapes anymore). GPS units were a hot Christmas item just three years ago. Today, they are just another free app for the iPhone. Even the camcorder has been completely replaced, first by digital recorders and later by mobile phone cameras.

Two years ago, Cisco Systems purchased the Flip digital camera business for a billion dollars, only to close it down as a complete loss.

Everything that can be made digital is becoming digital. An entire roomful of novels can now be replaced by a single eReader. A family photo album is now stored on a keychain.

This is part of a process that futurist Buckminster Fuller described as *ephemeralization*. It is the ability of technological advance to do "more and more until you can do absolutely everything with nothing." Buckminster Fuller's hope was that ephemeralization would provide endlessly improving standards of living despite global limits to physical resources.

We are also seeing an enormous level of convergence between electronic devices. An iPad is a book, Internet browser, music player, photo album, television, remote control, and game machine all in one sleek package.

The growing ease of wireless computing is also facilitating the possibility of mobile living. Pick up, pack up, and go.

When Big Boxes Shrink

If the outbreak of sharing continues, combined with the trends of *Less is More* and *Digitize Me*, we may see a lasting reduction in the need for more stuff as we manage resources more efficiently. Physical storage becomes less important as we learn to pool resources more effectively.

Smart retailers such as Amazon and GameStop are adjusting by selling both new and used products. They give consumers a choice on what they want and how much they are willing to pay.

Other companies, such as Plato's Closet, are focusing almost exclusively on the resale market. Stores are shifting slowly from retailing outlets to something that more closely resembles an exchange.

Our consumption habit is slowly going away. This is not necessarily bad. It is simply that we no longer need *Bigger, Faster, More* of everything -- we make the most of what we already have.

It is also about the shift away from the physical and toward the digital. The generation of students currently in high school and college has always lived in a digital world and their expectations may be different from previous generations. Ephemeralization may give our culture a huge amount of freedom in where we live and how we work.

In times of great change, expect a great amount of experimentation, with varying results. We'll likely see a much broader range of lifestyles as people try to figure out how to adjust. Not all income brackets will respond alike. The high end of the market may respond by *discreet consumption*. At the other end of the spectrum we may also see an uptick in *freegan* culture – those who live happily from society's cast-offs.

More Emerging Trends:

Discreet Consumption

For every action, there is an often an opposite reaction. While many people may continue to spend less by sharing more, the truly affluent may continue to overspend on basic items with enhanced performance and durability. Luxury markets never go away -- they sometimes just become invisible. Look for a growth in extremely high-quality basic goods with very subdued advertising and name recognition. These items will be marketed as bespoke or of professional quality.

A recent feature in Forbes profiled high-end fishing equipment. The list included a Tom Morgan Graphite Rod with "a traditional, almost bamboo-like feel, providing maximum accuracy" ($1,345), a Tibor QC Reel ($840), a pair of Simms G4 Pro Waders ($530), and a Patagonia Wading Jacket "the ne plus ultra of rain shells: waterproof, breathable, even fully recyclable" ($450).

Unlike conspicuous consumption, discreet consumption is about hiding how much you are spending. It's about spending significantly on what can be justified as practical and durable.

Dumpster Diving Divas

What happens when reality TV meets the grandfather of recessions? Competitive dumpster diving. It started with *Antiques Road Show* (PBS), devolved into *Pawn Kings* (History Channel), and hit bottom (almost) with *Hoarders* on A&E. Don't be surprised if television continues to scamper down the value chain as it explores freegan culture.

Vegetarian Chic

Fifteen years ago, it was almost impossible to find soy milk at the grocery store. Now you cannot only find soy milk, but also almond milk (multiple flavors), coconut milk, grass-fed cow milk, and just plain organic milk, as well.

Have you ever noticed that the vegetarian options on restaurant menus are almost always a few dollars cheaper than traditional meat-based entrees? There is a reason for that: It is simply more efficient to eat vegetables and grains than it is to eat the animals that are fed vegetables and grains. It takes 16 pounds of grain to make just one pound of beef. So the question is, should we be feeding people or cows?

This Tiny House

Radical reductions in consumption may lead to radically smaller new homes, many less than 800 square feet in size. Several dozen manufacturers have sprung up out of nowhere to support this growing movement over the past few years. These builders cover the range from yurts to log cabins, custom homes, and prefab units. "Tiny homes" are often small enough to be towed from one scenic location to the next, while still others can be mounted on the back of a pick-up truck.

The price is right, too. Many of these homes can be built for less than $30,000.

Online Resources:

BookCrossing: Places for booklovers to find, catch, and release books as they travel the world

www.bookcrossing.org

Couchsurfing: Worldwide network between travelers and homes

www.couchsurfing.org

Freecycle: Locally distributed free goods

www.freecycle.org

Neighborgoods: A place for neighbors to lend each other tools, sports equipment, appliances, etc.

www.neighborgoods.net

Read It Swap It: Matchmaking service for readers who want to swap books

www.readitswapit.com

Rehash: Trading community for clothing, books, and "coveted items"

www.rehashclothes.com

SharedEarth: Online matchmaker between landowners and gardeners

www.sharedearth.com

Swap: The largest online swap marketplace for books, DVDs, clothing, and games

www.swap.com

Swapstyle: The world's largest online fashion swap site

www.swapstyle.com

Thredup: Online exchange for used children's clothes and toys -- swapped by the box

www.thredup.com

Timebanks: Sharing platform for time and expertise within communities

www.timebanks.org

Zimride: Coordinate rides between coworkers and students

www.zimride.com

Tiny House Blog: Information on workshops, suppliers, and plans for tiny homes

www.Tinyhouseblog.com

Books:

Voluntary Simplicity by Duane Elgin

Your Money or Your Life by Vicki Robin and Joe Dominguez

Chapter 8

Radical Economics

Kyle MacDonald was just a guy with some ideas and a bunch of t-shirts. He was between jobs, living with his girlfriend. The t-shirts weren't helping to pay the rent, and his girlfriend was getting impatient with him.

On his desk was a stack of resumes, held together by a single red paperclip.

When he was a teenager, he and his friends played a game called Bigger and Better. The game begins by taking a small object to a neighbor and asking if he or she would be willing to trade something bigger or better for it. With some hard work and a little luck, it was possible to repeat the trade for bigger and better things.

There was a rumor going around that one kid started with a single toothpick and was eventually able to trade his way to a car. Kyle was wondering what he could get for a paperclip.

As a grown-up, Kyle didn't want to go around knocking on his neighbors' doors – that just seemed too pushy and uncomfortable. The goal was to make it easy and fun. His cousin suggested that he could post the offer on Craigslist, which had a section for barter trading. On his first trade, he exchanged his paperclip for a pen shaped like a fish. He also made some new friends.

On successive trades, he obtained a doorknob, and then a Coleman stove. With the Coleman stove, he offered to host a free BBQ dinner. That led to a power generator, a keg and an electric Budweiser sign, an instant party, a snowmobile, a trip, a moving van rental, a recording contract, a year of free rent in Phoenix, an afternoon with rock star Alice Cooper, and a KISS snow globe.

Going from a year's worth of free rent to a snow globe looked like a giant step backward. Most sane people would have stopped at the free rent. But this was a *motorized* snow globe, with multi-colored lights, and a variable speed dial. Possibly one of the finest snow globes ever made.

This led to an introduction to Corbin Bernsen, a film producer and avid collector of snow globes. In exchange for the snow globe, he offered Kyle a speaking role in a film. A movie role soon led to an offer for a home in Saskatchewan, Canada.

After a few months, Kyle traded his way from a red paperclip to a house.[1]

On the way, he met hundreds of people, became a local celebrity, and kept his girlfriend.

Kyle's story is a great example of the importance of networks. The people that Kyle met were far more important to him than the things that he received in his trades. If he hadn't made the right connections, he might have been stuck with a fish-shaped pen or a doorknob.

When people don't have access to cash, they sometimes get creative. People are cooperative by nature and in times of need, they are willing

---

[1] His last trade may have been better: Kyle later turned his story into a deal for a book called *One Red Paperclip*.

to work together to exchange whatever talents and assets they have available.

It took a post on Craigslist for Kyle to make his first trades. After that, he was able to advertise his trades on his blog. This helped him to get the word out and opened up a whole range of new possibilities.

As remarkable as this all sounds, Kyle was still practicing bartering 1.0, which relies on something called a "coincidence of wants." In other words, everyone might be willing to swap something for a haircut, but not all barbers want or need a snow globe. This is one of the reasons why money is so useful – it can be used in exchange for just about anything.

Bartering has since evolved to a new level. As an example, Swap.com now manages a huge network of booklovers who exchange copies of their previously read books. People can get more value by trading than they would if they sold their books at a second-hand store. Similar swaps are available for music, games, and DVDs – with shipping costs as the only expense.

Swap.com brings an added sophistication to bartering by arranging multiple simultaneous trades. So, if an owner of *Breaking Dawn* doesn't want a copy of *The Hunger Games*, but would settle for a copy of *Harry Potter*, and the owner of *Harry Potter* wants *The Hunger Games*, Swap.com will set up a three-way trade so that everyone gets what they want.

We'll call this the 2.0 level of bartering. At the end of the day, it is still a like-kind exchange of comparably equal value.

Bartering 3.0 is the establishment of bartering networks that enable the accumulation of wealth. Instead of needing to complete both sides of a transaction, one party can choose to receive payment in trade credits, which are deposited in an account. Later, when a business needs to buy something from any other member of the network, it can pay using these credits. These networks are known as LETS (local exchange trading systems).

Perhaps the oldest such network is the WIR Bank in Switzerland. Founded in 1934 as a result of currency shortages following the stock market collapse, the WIR Bank started as little more than a bookkeeping system to track the value of exchanges between local merchants.[2] By setting up as a trading system, participants were able to create buying power in the absence of cash. Economic activity continued to be recycled through the group, creating a "virtuous cycle" of transactions.

WIR Bank is a not-for-profit organization. It exists solely for the benefit of its members and provides a resilient alternative to conventional banks.

Almost any type of trading can be good for businesses; it doesn't matter whether services, goods, money, or credit is exchanged.

Bartering frees up needed cash for buying inventory, paying salaries, and reducing debt. What is good for cash flow is good for business, but just as important is the development of connections and relationships.

Futurist Heather Schlegel notes that there is often an increase in non-financial transactions during times of stress. "When you do these non-financial transactions, you experience a higher level of relationship with someone. Nonfinancial transactions create sticky relationships." When this happens within the context of a working community, networks of trust will develop over time. These networks play an important role in restoring confidence in the future. They represent potential customers, lenders, and word-of-mouth advertisers.

---

[2] *Wir* is an abbreviation for *wirtschaftsring* ("business circle") and also means "we" in German.

Many people don't own their own business, but they can still benefit from bartering. To follow Kyle's example, here is one way we can learn how to barter that is easy, social, and fun...

Food Swaps

Anyone who has grown a garden has had the experience of too much of *something*. Successful homesteaders often have an overabundance of their best crops.

There are only so many jars of jelly that can be consumed (or given away), so people are starting to host food swaps to trade and connect with like-minded fans of homegrown food. Food swaps happen when DIY and homesteading meet common sense.

Emily Ho organized the first food swaps in Los Angeles. She says:

> *A food swap not only gives members a chance to share delicious handmade foods but also is a wonderful opportunity to meet others who are interested in gardening, food preservation, beekeeping, and other sustainable DIY activities. As more and more people want to know where their food comes from and start activities like making their own condiments, baking bread, etc., it's fun to share this experience with others. Plus, who needs 20 jars of homemade ketchup? Better to swap and diversify your pantry!*[99]

With the assistance of websites such as Meetup and The Food Swap Network, many areas now have regular events where bakers, canners, farmers, homebrewers, and soap makers can exchange their goods. Swap events allow direct trades between participants on an item-by-item basis.

These are generally not exchanges of basic foodstuffs. One newspaper correspondent went to a food swap with a batch of traditional chocolate chip cookies and was overwhelmed at the inventiveness of the other offerings. At a Brooklyn event, she found such rarities as candied orange peels, four-fruit marmalade, raspberry marshmallows, and fig chutney.

There is a social component as well – some food swaps include potlucks where people can sample each other's goods before offering to make an exchange for more. In every regard, it is a good way of meeting other local producers and building a sense of community while enriching the variety of one's kitchen cupboard. One enthusiast has met so many people at these events that she refers to her local food swap as a "friend factory."

Food swaps provide a social way of introducing people with different needs and wants. The same type of approach also applies for more sophisticated economic transactions.

DIY Banking

Just as bartering reduces the need for cash, innovations in finance can diminish the need for conventional banks.

A bank typically serves as a middle man in financial transactions. They pay depositors a minimal amount of interest on their accounts and then lend the money at much higher levels of interest to their borrowers. The difference goes into the bank's profit.

What if people could find borrowers and make their own loans? They could earn much higher levels of interest than they could with a traditional bank, and have some control over where their money goes.

Prosper.com enables investors to do this through a website that matches prospective lenders and borrowers. Borrowers can select lenders based on their personal narratives, needs, or credit quality. All loans use a fixed payback period of three to five years. After the initial match, Prosper administrates each of the loans – including payment collection and record keeping. Prosper is not registered as a bank and loans made through the service do not receive the benefit of FDIC insurance.

So far, Prosper has matched $280 million in loans between over a million participants. Borrowers with excellent credit histories are currently paying over 5% interest, while at-risk borrowers looking to consolidate credit card debt are paying over 20%.

In order to control exposure to default risk, many lenders at Prosper set up diversified portfolios of loans. Individual loans can be for amounts as small as $25. This greatly minimizes exposure to loss in case any of the individual borrowers fail to make their monthly payments.

With Prosper and similar services such as Lending Club, banking becomes a more human experience. Peer-to-peer lending puts people in touch with what their money is doing. Borrowers, meanwhile, realize that their payments aren't going to some faceless corporation, but to real people who are depending on them to make their payments.

Financial innovation doesn't just stop there. When times are tough, people can be unusually resourceful at fundraising. Sometimes it takes a village of customers to help small businesses grow.

Deli owner Frank Tortoriello was looking to move to a larger store but soon found that his local bank would not lend him the $4,500 that he needed. Getting the expansion that his business needed took some creative financing. Tortoriello decided to print his own money. These could not be U.S. dollars – that would be forgery. He launched his own currency, called "deli-dollars", which any of his customers could buy for $8.

After the new location was opened, each deli-dollar could be redeemed for $10 of food. Tortoriello essentially paid off his loan in sandwiches while cultivating a loyal client base and creating free publicity.

"Frank's customers were backing his loan because they felt they were helping him beat the bank and he was paying them back in sandwiches," says David Boyle, an alternative economist who tells the story in his book, *Funny Money*.

It was not long before the deli-dollars took on a life of their own and found their way into general circulation. They were soon being used like real money.

"Parents passed them on to their student children to make sure they were eating properly," said Mr. Boyle. "Employers passed them to workers as Christmas gifts. The minister ate at the deli and soon notes started turning up in his collection box. Even the bank that refused Frank a loan in the first place circulated deli dollars."[100]

This was just the beginning...

Local Currencies

Going to the Berkshires in western Massachusetts is a bit like going to a foreign country. You won't need a passport, but they do have their own money. The Berkshare is a dollar-backed alternative currency used in the five-town area around Great Barrington. Local banks issue the money using an exchange rate of one Berkshare equals 95 cents. Once in circulation, Berkshares are treated just like U.S. dollars, but they can only be used to support local businesses. From a tax perspective, Berkshares share identical tax treatment as U.S. dollars.

Each denomination of the Berkshares features a different local hero, such as artist Norman Rockwell ($50), novelist Herman Melville ($20) and civil rights activist W.E.B. Du Bois ($5).

"The idea of local currency is local benefit," says Nick Kacher of the New Economics Institute. He also says by having strong local economies, communities partially shelter themselves from upheavals that occur on a national or global scale.

One local mainstay is the Red Lion Inn, which uses Berkshares to pay for produce from area farmers. The farmers, in turn, come back into town and spend their Berkshares at local stores.

For users who redeem U.S. dollars to buy Berkshares, it is similar to getting a 5% discount from any of 400 local merchants. Five towns around the Great Barrington area participate in the project, and thirteen bank branches do Berkshare exchanges. The money does not currently earn interest and there is a slight penalty for redeeming them for U.S. dollars. As a result, Berkshares tend to change hands quickly and remain in circulation.

Nick Kacher points out that because Berkshares are presently pegged to the value of the U.S. dollar, they will not provide protection if the dollar collapses in value. He says that the next step might be to peg the value of the Berkshare to a locally produced commodity, such as maple syrup. If ten Berkshares had a fixed exchange rate equal to a gallon of maple syrup, then they would still be worth a gallon of maple syrup, regardless of what happened to the U.S. dollar. If the value of a gallon of maple syrup went up, it would cost more U.S. dollars to buy Berkshares.

Other cities, such a Baltimore, Washington D.C., Philadelphia, New Orleans, and Detroit, have also created their own local currencies. Presently, there are over one hundred such efforts throughout the United States.

Paul Glover was largely responsible for the creation of the Ithaca Hours program. The original concept behind this currency was that each unit was worth one hour of time (roughly $10 U.S.), although locals are encouraged to negotiate the value of their Ithaca Hours. Professionals sometimes charge a multiple of Ithaca Hours to compensate for training and expertise, but many of them offer discounts to support the community.

The notes are colorful and printed on high-quality papers from cotton, hemp, or cattail fiber. Each bill uses a thermal ink that disappears when exposed to heat from photocopiers. The reverse side says, "This note entitles the bearer to receive one hour of labor or its negotiated value in goods and services. Please accept it, then spend it."

Ithaca Hours have changed the community in many interesting ways. Now, when visitors arrive from out of town, their local children have started to ask, "What does your city's money look like?"

Time Banking

Fifteen years ago, budget cuts were eliminating the funding for government social programs, creating shortages in basic human services. A Washington, D.C. area lawyer, Edgar Cahn, came up with an unconventional idea. Why not create a system that uses volunteer time as a type of currency? The people who have the least money sometimes have the most time. Using this approach, people could accumulate hours of service by contributing skills to each other and their community.

They can then "spend" these hours any way that they please, drawing from a pool of skills within their area. The amount of benefit received would equal the amount of time donated. Participants can search through a database of local talent and find the services that they need.

Time Banks USA provides a software infrastructure that enables communities to establish and administer their own time banking systems.

Stella Osorojos helped co-found a successful time bank in Santa Fe. She says, "Because time banking exists alongside a cash economy, where we pay money for our plumbers, dentists, and accountants, people turn to the Time Bank for the smaller interactions that make us neighbors, friends, and ultimately communities." This might include tasks like walking the dog, picking up kids from soccer practice, or cooking for someone who is sick. "These types of trades used to happen all the time and they forged bonds that were valuable because they were necessary to get along."[101]

In a cash-only economy, we mistakenly replace human interactions with financial transactions. This has the effect of creating a culture of isolation. When the economy was more robust, people used caterers for parties and landscapers for shoveling snow from driveways. Personal fitness trainers replaced friends at the gym and professional dog groomers took over the responsibility from the kid next door.

We can often forget the value of social ties and sometimes don't even take the time to learn what our neighbors have to offer. Time banking breaks down some of these barriers and encourages people to meet and connect with other people in their community.

Time banks are also incubators for small businesses. Volunteering as part of a time bank is a great way to get the word out and see if there is interest and demand from the community. They provide a lower-pressure way for people to cultivate their skills and develop contact lists before they open doors to new customers.

As with any grassroots effort, time banks can adjust to fit regional needs. In Japan, there is a strong tradition of assisting elderly parents. This has become a more difficult practice to maintain as families started to move apart due to the demands of work. In response to this growing problem, the Sawayaka Welfare Foundation developed a type of time banking by which a person could earn credits for taking care of an elderly person in their neighborhood and then apply those credits toward the care of one's parents in a different part of the country. This system is available throughout the country – a simple software system maintains all of the necessary records.

Time banking and related programs are not a cure-all, but they do provide a better way for people to work together and provide mutual assistance. In most ways, they could be regarded as complementary to our existing money-based system, reducing the need for financial transactions but not eliminating them altogether.

Rent a Life

There are other signs that we may slowly be moving away from our dependence on the traditional banking system. Banks may find that customers are becoming increasingly reluctant to take on additional debts when the future of the economy is so uncertain. Banks themselves have gotten much more cautious with their lending. These days, if you qualify for a personal loan, you might not actually need it.

The trend *away* from mass consumption will have some interesting implications for banks and credit cards.

One question that comes to mind is "what happens when people stop buying stuff?" Also, "what happens when economic thinking shifts away from the concept of ownership?"

There are three problems with ownership in the traditional sense:

- **Cost:** Ownership typically involves an up-front cost equal to the value of the item. Alternatively, financing larger purchases creates a long-term commitment of cash flow.
- **Idle Capacity:** Most stuff is used just a fraction of the time. We keep things because we might need them again sometime in the future. Meanwhile, they just take up space.

- **Lock-In:** Ownership makes us to choose what we want, even if it is not always the ideal choice. Ownership leads us toward a lifestyle of compromise and inflexibility.

Cars provide a great example of why people are making the shift from ownership to access. Rachel Botsman, author of *What's Mine is Yours,* says, "The average car costs eight thousand dollars a year to run. Yet that car sits idle for twenty-three hours a day. So when you consider these two facts it starts to make a little less sense to own one outright."[102]

Hourly rentals from Zipcar are revolutionizing the car industry. Zipcar gives people freedom and flexibility. Why limit yourself to owning an all-purpose minivan when what you really want is a convertible for Friday night and a pick-up truck for running errands on Saturday morning? Zipcar allows you to match your car to the occasion.

Furthermore, if you are traveling out of town and not using it, you don't have to pay for parking, insurance, or car payments.

Younger consumers are beginning to shift away from owning cars to renting them when needed – sometimes from their friends and neighbors. Rachel Botsman refers to this trend as "collaborative consumption."

Peer-to-peer rentals enable car owners to earn money on their vehicles when they don't need them. Renting out an unused car for just 15 hours per week can provide enough cash flow to cover a monthly loan payment.

Companies like Getaround, RelayRides, Go-Op, and Jolly Wheels connect people who own cars with people who need them.

Getaround provides a smartphone app, which provides maps and directions to available cars nearby. The app also unlocks vehicles for customer use. In case something goes wrong, Getaround has insurance coverage on all rentals. At the end of each transaction, renters and owners are encouraged to rate their experiences with each other. These ratings and comments are viewable by other users.

Peer-to-peer rentals are branching out into a mind-boggling range of areas. Online ventures such as RentMineOnline, Zilok, and iRent2you enable people to rent just about anything to each other. Current listings include offers for consumer electronics, sporting goods, and evening gowns. AirBnB created a market for people to rent out spare rooms and even empty sofas to travelers. Today, over 160,000 people in 150+ countries are part-time innkeepers on AirBnB. This makes AirBnB one of the fastest-growing hospitality companies in the world.

Peer-to-peer rentals only make sense when there are sufficient opportunities and population density. As such, this lifestyle seems to be gaining the most traction in urban areas such as New York, Boston, and San Francisco. There is a network effect involved here: The more people do it, the more convenient and common it will become.

If the Millennials are the first "post-consumer" generation, then we could expect to see many more peer-to-peer transactions in the future.

So what are the implications?

- This is all about the ephemeralization of our economy. It is about doing more with less.
- We are learning how to be more efficient by becoming more social. This enables us to maintain our standards of living while minimizing our resource needs.
- The trend away from mass consumption lends itself to greater personal freedom and flexibility. People may be less likely to become "locked in" to their lifestyles and more able to adjust their expenditures to reflect changes in income.
- For the past few decades, people have relied on their financial assets to generate cash flow. In the emerging culture, people

may be more likely to generate cash flows from their physical assets to help cover the expense of ownership.

- Consumer loans for cars and other large purchases may become less common, as people become more hesitant to take on long-term financial commitments.

Banks will never go away completely; in a more resilient world we will simply need them *less*.

Online Resources:

Complementary Currency Resource Center: History, research, and support for alternative currencies

www.complementarycurrency.org/materials.php

Food Swap Network: Details on how to host or find a food swap

www.foodswapnetwork.com

IRent2u and Zilok: Online rental marketplace connects local renters to owners

www.irent2u.com

www.zilok.com

Kickstarter: Crowd-sourced fundraising for artistic endeavors

www.kickstarter.com

Lending Club and Prosper: Peer-to-peer borrowing and lending

www.lendingclub.com

www.prosper.com

Swap: Free trading of books, DVD's, and music

www.swap.com

Time Banks USA: Education and resources for establishing local time banks

www.timebanks.org

Trade Smart Now and Barter Connections: Internet bartering exchanges for businesses

www.tradesmarketnow.com

www.barterconnections.com

Zilok: Online peer-to-peer marketplace for renting anything

www.zilok.com

Zipcar: Hourly car rentals in dozens of cities

www.zipcar.com

Zopa: Peer-to-peer lending in the U.K.

www.zopa.com

Chapter 9

Working It

Justin Caggiano is a laid-back rock-climbing guide whom my wife and I met during our last vacation in the red canyons of Moab, Utah. He's also been guiding rafters, climbers, and hikers for the past six years.

We watched Justin scramble up the side of a hundred-foot natural wall called The Ice Cream Parlor, a nearby climbing destination that earned its name from remaining shaded and cool in the morning, despite the surrounding desert. His wiry frame allowed him to navigate the canyon cliffs and set up the safety ropes in a fraction of the time that it took us to make the same climb later that day.

Justin's rock-climbing skills easily translated into work as an arborist during the off-season, climbing up trees and then cutting them from the top down to prevent damage to nearby buildings. Since graduating from college six years ago, he has also worked as an artisanal baker, a carpenter, and a house painter. This makes him something of a down-to-earth renaissance man.

His advice is, "Be as flexible as you can -- and work your tail off."

It is an itinerant lifestyle for Justin, who frequently changes his location based on the season, work, and nearby climbing opportunities. Rather than committing to a single employer, he pieces together jobs wherever he can find them. His easy-going personality enables him to connect with people and find new opportunities when they become available.

This winter, he plans to stay with a friend who is building a house. Justin is planning to help out with carpentry and wiring in exchange for free rent.

He's been living on shoestring for a while now, putting away money every year. Longer-term, he'd like to develop all of the skills that he needs to build his own home and then pay for land and materials entirely with savings from his bank account. He plans to grow fruit trees and become somewhat self-sufficient. After that time, he says, "I'll work when I'm needed and live the debt-free, low-cost lifestyle when I'm older."

There is a difference between earning a living and making a life.

Clearly, our concept of work is getting reworked. A career used to be a ladder of opportunities within a single company. For the post-war generation the concept of "lifetime employment" was a realistic expectation. My father worked 33 years at DuPont as a research scientist and spent almost all of that time at a sprawling complex called the Experimental Station. Most of my friends' parents had similar careers. Over time, they were gradually promoted and moved up the ladder. At best, it was a steady progression. At worst, they found their careers stuck in neutral.

The Boomers had a somewhat different career trajectory. They still managed to have a single career, but it more closely resembled a lattice than a ladder. After working for an employer for 5-10 years, they might find a better opportunity elsewhere and continue their climb. The successful ones cultivated networks at related businesses and continually found better opportunities for themselves.

The career path for younger generations more closely resembles a patchwork crazy quilt, as people attempt to stitch multiple jobs together into something that is flexible and works for them. In today's environment, they sometimes can't find a single job that pays enough to cover all of their expenses, so like Justin, they find themselves working multiple jobs simultaneously. Some of these jobs might match and be complementary to existing skills, while others can be completely unrelated.

Multi-Tasking Careers

Pamela Slim, author of *Escape from Cubicle Nation*, encourages corporate employees to start a "side hustle" to try out new business ideas. This would provide another source of income to supplement an existing job or business. She also recommends having a side hustle as a backup plan in the event of job loss. This strategy is not just for corporate types; Slim says, "It can also be a great backup for small business owners affected by shifting markets and slow sales."

She says that an ideal side hustle is a moneymaking activity that is doable, enjoyable, can generate quick cash flow, and does not require significant investment. Examples that she includes are businesses such as web design, massage, tax preparation, photography, and personal training.

The new norm is for people to maintain and develop skill sets in *multiple simultaneous careers*. In this environment, the ability to learn is something of a survival skill. Education never stops, and the line between working and learning becomes increasingly blurred.

After getting her Ph.D. in gastrointestinal medicine, Helen Samson-Mullen spent years working for a pharmaceutical company, first as a medical researcher and then as an independent consultant. More recently, she has been getting certifications for her career transition as a life coach. Clinical project management is now her "side hustle" to bring in cash flow while she builds her coaching business. Meanwhile, she is also writing a book and managing her own website. Even with so many things happening at once, Helen says, "My life is so much less crazy now than it was when I was consulting. I was always searching for life balance and now feel like I'm moving into harmony."

Her husband Rob is managing some interesting career shifts of his own and is making a lateral move from a 22-year career in pharmaceuticals to starting his own insurance agency with State Farm.

Fixed hours, fixed location, and fixed jobs are quickly becoming a thing of the past for many industries, as opportunities become more fluid and transient. The 40-hour workweek is becoming less relevant as we see more subcontractors, temps, freelancers, and self-employed. The U.S. Government Accountability Office estimates that these "contingent workers" now make up a third of the workforce. Uncertain economics make long-term employment contracts less realistic, while improvements in communications make it easier to sub-contract even complex jobs to knowledge workers, who login from airports, home offices, and coffee shops.

Mind the (Generation) Gap

The "good old days" of lifetime employment with a secure pension are gone for good. The rate of change in the workplace has been almost exponential during the past few decades. This is perhaps best reflected in generational perspectives toward work. In order to really understand these changes, it is useful to consider the shifting hopes and expectations of various generations as they enter the workplace.

Twenty years ago, Generation X graduated during what was a comparatively soft recession in the early 1990s. The so-called Thirteenth Generation was branded by the media as being "cynical, directionless, and apathetic." Subversive films such as *Clerks, Reality Bites*, and the aptly titled *Slacker* did nothing to change that reputation.

In comparison, the Millennials are making members of Generation X look like career-obsessed workaholics.

The Millennial generation faces the dual challenge of a weak labor market combined with competitive (and rising) educational standards. It's been over fifty years since so many young people in America were out of work. The American Dream of home ownership and 2.5 kids is beginning to look like a bum deal. Why have equity in a sinking asset? In such an uncertain world, does having children even make sense? Census data show that people are getting married later in life – if at all.

To get a sense of where things are going, it is useful to look at the average Millennial. Every bit of personal history – photos, music, favorite movies, books – can be kept on a laptop. The new status symbols for good jobs are freedom, mobility, and fun. The key to attracting Millennial employees is to give them work that is engaging and the ability to connect with other interesting people. Financial compensation is sometimes not as important as freedom to choose when, where, and how work gets done.

Why can it be so hard to motivate this generation financially?

Here is a clue: A survey by Pew Foundation reveals that one in eight Americans aged 22-29 have "boomeranged" back to living with their parents after living on their own. (There is more on this to follow in the chapter on Family.) Many others are sharing living spaces with their peers to save money. Car ownership is actually declining somewhat, in favor of public transportation or services such as Zipcar.

It appears that the Millennials have been the first generation to fully adjust their long-term economic expectations. They have figured out that "access" often makes more sense than "ownership", particularly since "ownership" often translates into "debt".

The Millennials aren't unmotivated — they simply haven't bought into the old economic system.

Resilience and the Future of Everyday Life

Sarah recently shared her story and posted her comments on my blog:

From a very young age, while playing the game of LIFE, I never wanted to go into debt. It scared me. I was afraid that I would never be able to pay it back - how could I make sure that I made enough money to pay it back?

I think that my generation understands the instability of life; we have seen a lot of things that have changed the way the generations before us operated in the world. Granted, we didn't live through the Great Depression or WWII or Vietnam, but we see similarities in what is happening around us - the recession, the multiple wars overseas, climate change, terrorism, etc.

I did go to college. I moved from a great career-oriented job in a big city to a small company with great flexibility in a college town. My younger brother went to college (for finance). He has boomeranged back home, had a job that he quit because it wasn't what he wanted and they were going to transfer him. He is still living at home, trying to find work. A girlfriend of mine lived at home while going to college and just bought the house right next to her parents. A few of my girlfriends from college are all living together in a house owned by one of their parents; they all work in health care.

The common theme for all of us is that we know what we want. Or, more accurately, we know what we don't want. We don't want to be overworked and miserable like we have seen our parents/older siblings/other family. We want to maintain our independence so that we can travel, party, and play the real game of life.

In some regards, the Millennials may be the first post-consumer generation. This changes how they buy things (or not) and why they take on new jobs (or not).   In all likelihood, this may change somewhat as the Millennials eventually settle down and start families of their own. They may follow the Generation X habits of do-it-yourself projects and homesteading, or they may create entirely new forms of shared jobs and ownership.

To understand the past, present, and future of work, it is very useful to look at how different generations think about their careers:

Categories	Silents/Veterans/ Traditionalists	Baby Boomers	Generation X	Generation Y /Millennials
Birth Dates	1925-1945	1946-1960	1960-1980	1981-2001
Population	55 million	76 million	60 million	74 million
Expected Education	High school diploma	College degree	College plus some graduate education	Lifelong learning: no rush to start or finish college
Role Of Career	Means for living

1 or 2 career positions | Central focus

6 career positions | Irritant

12 *different* careers | Always changing

20 *different* careers |
| View of Authority | Honor and respect for leaders | Challenge leaders | Don't try to become one | Respect authority, but not awed by it.

Defer to a |

				team.
View of Technology	Hope to outlive it	Master it	Enjoy it	Employ it
View of Success	Fought hard and won it	Were born to have it	Successful because they're adaptable	Successful because they're tenacious
Interactive Style	Team player	Self-absorbed	Entrepreneur	Team player
Motivational Messages	"Your experience is valued here." "Your perseverance is valued and will be rewarded."	"You're important to our success." "We recognize your unique contribution to our team."	"Do it your way." "We're not very Corporate." "We've got the latest technology."	"You'll be working with other bright, creative people." "You are making a positive difference to our company."

(source: Courtney L. Via, Ph.D., University of Phoenix Knowledge Network)

While Generation X challenged the status quo of "business as usual", the Millennials have seemingly abandoned it altogether.

In many ways, this makes the Millennials particularly well suited for work in the emerging culture.   They've adjusted their expectations of the workplace and have a level of open-mindedness and flexibility that we've never seen before.   They have also learned how to manage downtime -- if there is not already enough work to keep them busy, Millennials just add on more freelance jobs, take additional graduate school courses, or spend the time training for a different career altogether.

Work Locally, Compete Globally

There is an increasing break between the digital and the real. In the real world, which provides physical goods and physical services, we may see a continuation of "business as usual." This means that for a large part of the workforce, there will be a continuation of the 40-hour workweek and regular commutes to the office.

Workplaces that require the maintenance of physical facilities or equipment will continue to need people in traditional jobs, because flexible locations and hours do not work well for everything.

For example, factory equipment and large office spaces cannot be easily moved. Having employees at the cash register is a good idea whenever there are people in the store. In an ideal world, baggage handlers should be available just prior to a flight. Few people would disagree that mowing lawns or painting houses is best done on-site and during daylight hours.

If anything, the trend toward flextime and location seems to be most relevant to the knowledge workers and other people who deal with the flow of information. This would include programmers, writers, design people, and anyone involved in communication. Increasingly, it will also involve teachers, entrepreneurs, engineers, lawyers, and physicians.

37 Signals is a software development firm headquartered in Chicago, but more than half of its team lives elsewhere. They have staff living in Spain, Canada, Idaho, Oklahoma, and elsewhere. They still make sure that the whole team meets together a few times a year. Founder Jason Fried says, "Geography just doesn't matter anymore. Hire the best talent, regardless of where it is."

This is creating not only a sense of global competition for work, but also a type of new mobility. It is increasingly possible for the most capable knowledge workers to choose not only when they work, but how they work. This creates the freedom live in places that are more affordable or more personally rewarding. It is just another indicator that we'll be seeing more in the way of nomadic lifestyles.

Some people are choosing to live abroad, simply because they can. Patrick Dugan is a game designer who now works out of Buenos Aires. He moved in 2007 because it seemed like a good place to set up a business. The cost of living was lower in Argentina and he was able to find talented people to write code for a fraction of the amount of what it would cost in the U.S. On a personal level, he found the culture more vibrant and the dating scene more lively than it was back home.

Things have changed over the past five years for Dugan. Since his original move, the cost of living in Argentina has increased by fifty to sixty percent. He now has a serious girlfriend and supports two kids. In considering his current options, he is thinking about relocating to Brazil or Uruguay.

In the past, people would hop from one job to another. Today, smart people are freer than ever to hop on a plane and relocate to a different continent when their personal needs or preferences change.

Where are the jobs?

Jobs are not only being outsourced to people in other countries, they are being *othersourced* to automated workers. Jared Weiner, a futurist and consultant at Weiner Edrich Brown, notes that we'll see more white-collar jobs lost to software algorithms and intelligent computers over the coming years. While automation has already had a significant impact on manufacturing, we are just beginning to see the impact of artificial intelligence on the traditional professions.

He notes that the financial services industry is becoming increasingly othersourced and is experiencing a modern industrial revolution of its own. According to Weiner, "Those jobs are not going to return – they can be done more efficiently and error-free by intelligent software."

In the investment business, we are seeing the replacement of financial analysts with quantitative analytic systems, and floor traders with trading algorithms. Mutual funds and traditional portfolio managers now compete against ETFs (exchange-traded funds), many of which offer completely automated strategies.

Industries that undergo this transformation don't disappear, but the number of jobs that they support changes drastically. Consider the business of farming, which employed half the population in the early 1900s but it now provides just 3% of all jobs. The U.S. is still a huge exporter of food; it is simply far more efficient now in terms total output per farm worker.

In an ideal world, jobs would be plentiful, competitive, and pay well. Most job opportunities have two of these qualities, but not all three. There are some jobs that are competitive and pay well (medicine, law, finance). Or there can be many jobs that are competitive but pay low wages (retail, hospitality, personal services). Unions often ensure that jobs pay well and are plentiful, only to later find that those jobs and related industries are no longer competitive.

Since 1970, manufacturing jobs, as a percentage of total employment, have declined from a quarter of payrolls to less than ten percent. Some of this decline is from outsourcing; some is a result of othersourcing. Those looking for a rebound in manufacturing jobs will likely be disappointed. These jobs will probably not be replaced – not in the U.S. and possibly not overseas, either.

This is all a part of the transition toward a post-industrial economy.

Jeff Dachis, Internet consulting legend and founder of Razorfish, coined the phrase "everything that can be digital, will be." To the extent that the world becomes more digital, it will also become more global. To the extent that the economy remains physical, business may become more local.

Work Will Make Us More Human

David Autor is an economist at MIT who has developed some useful insights on the impacts of outsourcing and othersourcing. According to Autor, the jobs that are currently being lost involve middle-skilled cognitive and productive activities. These tasks follow clear and easily understood procedures that can reliably be transcribed into software instructions or subcontracted to overseas labor. Autor writes that labor markets worldwide are rapidly becoming polarized, and he sees a clustering of job opportunities at opposite ends of the skills spectrum.[103]

At one end of the spectrum are low-paying service-oriented jobs that require personal interaction and the manipulation of machinery in unpredictable environments. Examples might include driving a vehicle in traffic, cooking food in a busy kitchen, or taking care of cranky preschoolers. Unless people decide to freight their toddlers to India for cheaper childcare, these tasks will still need to be performed locally.

Personal judgment and common sense can be important in even the most basic service jobs. "The added value of the worker is during the non-routine parts," says Autor. This might include a truck driver navigating an eighteen-wheeler through road construction or a security guard identifying suspicious activity at a department store.

To the extent that many service jobs involve human interaction, they also require skills such as empathy and interpersonal communication. Good employees can see things from the perspective of their customers.

At the other end of the spectrum are jobs that require creativity, ambiguity, and high levels of personal training and judgment. These jobs tend to pay well, because they require skill sets that are more difficult to replicate.

The jobs that are disappearing are those that are dirty, dull, or dangerous. These jobs are best left to automation. In fact, the root of the word "robotics" can be traced to a Czech word meaning "drudgery".

The job opportunities of the future require either high cognitive skills, or well-developed personal skills and common sense. In a nutshell, people will need to be either "smart" or "nice" to be successful (preferably both!)

Luddites should take notice – computers just might push us to do work that is meaningful and enables us to become better people. The activities that make us human – thinking, dreaming, learning, communicating, and feeling – are the skills that are the most difficult to program. In a contest of "man vs. machine," people will continue to shine and outperform in these areas for years to come.

Perhaps just as important are the significant advantages that people have in developing and maintaining relationships with others.

Reputational Marketing

In the emerging economy, the focus will no longer be just on producing more for less. During a credit crunch, demand is reduced along with production. As such, there won't be factories running at one hundred percent capacity. Managers might not need sixty-hour workweeks anymore. The focus shifts from working hard to working well. It is a focus on maintaining the quality of the product and building on professional reputation, knowing that customers are not just buying a product; they are making an investment in something that they expect will last them for a long time. The beauty of this, of course, is that we can take pride in our work again. Lack of time is no longer an excuse for poor quality. The winners of the restructuring will be those with the best reputations.

The significance of reputation is magnified by the presence of the Internet. While searching the Internet was initially an exercise in finding the best deal, it is rapidly becoming a means of comparisons based on reputations and consumer reviews. Amazon.com almost single-handedly put many bookstores out of business based on price, but user reviews are now the reason many users go there to research products before buying.

Consumers can sing their praises (or broadcast their complaints) not only through social networking sites (Facebook, Twitter, Google +), but on sites that specialize in rating businesses and local services (Angie's List, Yelp!, Yahoo Local). This goes far beyond just rating service at the local coffee shop. Angie's List now rates over five hundred different types of service providers, including plumbers, roofers, auto mechanics, landscapers and locksmiths.

"People are going to search out and see who you are before they do business with you," says Joyce Gioia, a management consultant at the Herman Group.

Traditional white-collar professions are subject to online scrutiny, as well. There are at least a dozen online sites where doctors can be anonymously reviewed and rated by former patients. Similar sites exist for financial advisors and lawyers, too. While word-of-mouth advertising traditionally occurred between friends, reputations are now most often shared over the Internet between complete strangers.

In the past year or two, there has been a counter-surge in companies that specialize in "reputation management." These web consultants can help companies suppress negative search findings and improve the ranking of searches that show them in a good light. Hollis Thomases, president of Web Ad.vantage (a digital marketing firm in Havre de

Grace, Maryland) says that in the final analysis, "companies need to mind their P's & Q's, because the real stuff is going to carry a lot more weight."

Fearful businesses have responded to the slowing economy by instituting strategies that are marginally helpful in the short-term but ultimately detract from customer satisfaction. Thomases uses the examples of banks that add extra fees on credit cards and airlines that cut legroom for passengers or add costs to check in luggage at the gate. All of these things may be good for the bottom line but damaging to reputation.

In a slower-growth economy, smart companies will realize that lavishing attention on their customers is a highly effective use of time. Rather than irritating clients with new user fees or restrictive policies, companies can use the downtime to ensure that their customers will say great things about them.

The Future of Work

Work will always be about people. It is about finding what other people want and need -- and then creating practical solutions to fulfill those desires.

What is changing right now is all of our basic assumptions about how work gets done. It is less about having a fixed location and schedule and more about thoughtful and engaged activity. Increasingly, this inspiration can happen anytime, anyplace.

There is a blurring of distinctions between work, play, and professional development. The ways that we measure productivity will be less based on time spent and more based upon the value of the ideas and the quality of the output. People are going to have a much better awareness of when good work is being done.

The fixed-time and location-based jobs will continue to exist, of course, but these will become smaller slices of the overall economy.

There are some tradeoffs involved. The old model of work provided an enormous level of predictability. In previous eras, people had a sense of job security and knew how much they would earn on a monthly basis. This gave people a certain sense of confidence in their ability to maintain large amounts of debt. Our consumer economy thrived on this system for more than half a century.

The new trends for the workplace have significantly less built-in certainty. We will all need to rethink, redefine, and broaden our sources of economic security. To the extent that people are developing a broader range of skills, we will also become more resilient and capable of adapting to change.

Finally, we can expect that people will redefine what they truly need in a physical sense and find better ways of fulfilling their needs. This involves sharing and making smarter use of the assets we already have. Meanwhile, businesses are doing the same. The outcome could be an economy that balances the needs between economic efficiency and human values.

More Emerging Trends:

Results-Only Workplace Environments

Imagine an office where meetings are optional. Nobody talks about how many hours they worked last week. People have an unlimited amount of vacation and paid time off. Work is done anytime and anywhere, based entirely on individual needs and preferences. Finally, employees at all levels are encouraged to stop doing anything that is a waste of their time, their customer's time, or the company's time.

There is a catch – quality work needs to be completed on schedule and within budget.

Sound like a radical utopia? These are all basic principles of the Results Only Work Environment (ROWE), as pioneered by Cali Ressler and Jody Thompson while they were human resource managers for Best Buy.

It is "management by objective" taken to a whole new level.

Best Buy's headquarters was one of the first offices to implement the ROWE a little over five years ago. The movement is small but growing. The Gap Outlet, Valspar, and a number of Minneapolis-based municipal departments have implemented the strategy. Today, ten thousand employees now work in some form of ROWE.

Employees don't even know if they are working fewer hours (they no longer count them), but firms that have adapted the practice have often shown significant improvements in productivity.

Cali Ressler says, "Thanks to ROWE, people at Best Buy are happier with their lives and their work." The company has benefited too, with increases in productivity averaging 35% and sharp decreases in voluntary turnover rates, as much as 90% in some divisions.[104] Interestingly enough, the process tends to reveal workers who do not produce results, causing involuntary terminations to creep upward.

ROWE managers learn how to treat their employees like responsible grownups. There is no time tracking or micro-management.

"The funny thing is that once employees experience a ROWE they don't want to work any other way," Ressler says. "So employees give back. They get smarter about their work because they want to make sure they get results. They know that if they can deliver results then in exchange they will get trust and control over their time." [105]

Co-Working

There are new alternatives to either working at home alone or being part of a much larger office. Co-working spaces are shared work facilities where people can get together in an office-like environment while telecommuting or starting up new businesses.

"We provide space and opportunity for people that don't have it," says Wes Garnett, founder of the coIN Loft, a co-working space in Wilmington, Delaware.

Getting office space in the traditional sense can be an expensive proposition, with multi-year leases, renovation costs, and monthly utilities. "For $200 (a month), you can have access to presentation facilities, a conference room, and a dedicated place to work." The coIN Loft also offers day rates for people with less frequent space needs.

According to Garnett, more people are going to co-working spaces as "community centers for people with ideas and entrepreneurial inclinations." He explains that co-working spaces provide a physical proximity that allows people to develop natural networks and exchange ideas on projects.

"We all know that we're happier and more productive together than alone" is the motto for nearby Independents Hall in Philadelphia.

Co-working visas enable people to choose from among 200 locations across the U.S. and three dozen other countries.

Silicon Colleagues

Expert systems such as IBM's Watson are now "smarter" than real people – at least on the game show *Jeopardy*. It was a moment in television history when Watson decimated previous human champions Ken Jennings and Brad Rutter on trivia games, which included categories such as "Chicks Dig Me."

IBM's Watson is a software-based knowledge system with unusually robust voice recognition. IBM has stated that its initial markets for the technology are healthcare, financial services, and customer relations. In the beginning, these systems will work side by side with human agents, whispering in their ear to prompt them with appropriate questions and answers that they might not have considered otherwise. In the next decade, they may replace people altogether in jobs that require simple requests for information.

"It's a way for America to get back its call centers," says futurist Garry Golden. He sees such expert systems reaching the workplace in the next two to three years.

Opting-Out

A changing economy is causing people to rethink their priorities. In a recent survey by Ogilvy and Mather, 76% of respondents reported that they would rather spend more time with their families than make more money. [106] Similarly, the Associated Press reported that less than half of all American reported that they were happy with their Jobs.[107]

Given the stresses of the modern workplace, it is not surprising that more people are simply "opting-out" of the workforce. Since 1998, there has been a slight decline in the labor force participation rate – about five percent for men and three percent for women. This trend may accelerate once extensions to unemployment benefits expire. Some of these people are joining the DIY movement, and others are becoming homesteaders.

To a certain extent, this reflects a shift back toward one-income households that happens when the cost of taxes, commuting and childcare consume a large portion of earnings. People who opt-out are not considered unemployed, as they are no longer actively looking for paid work. Their focus often reflects a shift in values toward other activities, such as raising kids, volunteer work, or living simply.

Online Resources:

Career Diva: Career advice, labor issues, job news, and opportunities

www.careerdiva.net

Co-working Visas: Locate co-working facilities around the world

www.wiki.coworking.info/w/page/16583744/CoworkingVisa

Entrepreneur: Tips for self-employed and small business owner

www.entrepreneur.com

Indeed: Online job search engine -- aggregates data from thousands of career sites and job boards

www.indeed.com

Freelancers Union: An online support system for independent workers -- access to insurance and retirement benefits

www.freelancersunion.org

LinkedIn: Professional-networking site with access to job postings

www.linkedin.com

Resources to Help Balance Work, Life, and Family:

www.humanresources.about.com/od/worklifebalance/Resources to Help Balance Work Life and Family Employee Assistance.htm

Small Business Administration: How-to info on small business start-ups and financing www.sba.gov

Books:

Career Distinction: Stand Out by Building Your Brand by William Arruda and Kirsten Dixson

Coach Yourself to a New Career: Seven Steps to Reinventing Your Professional Life by Talane Miedaner

Why Work Sucks by Cali Ressler and Jody Thompson

Rework by Jason Fried and David Hansson

Escape from Cubicle Nation, by Pam Slim

The Mind at Work: Valuing the Intelligence of the American Worker by Mike Rose

Chapter 10

Post-Nuclear Family

Have you ever looked at your life as a sitcom? It seems like everyone has crazy uncles, quirky neighbors, and precocious children. When Steve Levitan and Christopher Lloyd looked at their own families, they realized that they had plenty of material. Sharing their personal backgrounds gave them the missing ingredients to update the formula for dysfunctional domestic sitcoms. The result was the Emmy award-winning show, *Modern Family.*

When art imitates life, it is called realism. When comedy imitates life, it makes us laugh.

I must confess that I am a bit behind the times when it comes to television. My wife and I haven't had one in our house in over fifteen years. However, since the advent of Hulu and Netflix, it is no longer necessary to have a TV to get "plugged in."

What makes *Modern Family* so fascinating to watch is that the producers completely understand family dynamics in the early 21st century. It is microcosm of the changes we are seeing today.

Let's start with the show's traditional nuclear family. There is Phil, who attempts to be the "cool dad." In the show's first episode he brags, "I'm hip, I surf the Web, I text. LOL: laugh out loud, OMG: oh my God, WTF:

why the face?" Phil desperately tries to fit in with his kids and spends equal time attempting to placate his neurotic and assertive wife, Claire.

Phil isn't the only one who has difficulty acting his age. Claire's father, Jay, is a rich old guy with a fabulous house and a too-hot-to-handle younger wife from Columbia. Meanwhile, Claire's brother Mitchell is in a same-sex relationship with Cam; they share an adopted baby girl from Vietnam.

The characters in *Modern Family* don't talk *to* each other as much as they talk *around* each other – via e-mail, Skype, and Facebook. Instead of shouting up the staircase, they send each other text messages. Technology has taken up residence as part of the cast.

From this perspective, the family of the 21st century is bigger, more colorful, a bit more openly gay, and substantially more confused.

The message of the show is that while there are always plenty of opportunities for miscommunication and differing opinions, "no problem is too big that it can't be swept under a hug."[108] Tolerance of sexual orientation and racial diversity is something of a survival skill; without it, there would be no modern family.

In the show, family members are often wandering in different directions, but there is still a core of kinship that keeps them together. As Jay says, "Ninety percent of being a dad is just showing up."

For an institution that has been around for a longer period of time than our species, it is remarkable to watch the rate of change that has occurred in family structure over the past century. We've gone from a period of extended families, to nuclear families, to something that could only be described as *post-nuclear families*.

Post-nuclear family is still family, but mutated into dozens of different forms. The nuclear family will continue to be the most common and basic structure, but we're going to see more and more variations over the course of the next decade. Some of these family structures are very old, while others will be entirely new. We'll also start to see more *mix-and-match* families, as people pull together resources for mutual benefit.

Re-Extended Families

Close to seventeen percent of all families in the U.S. now live in extended family households containing at least two generations of adults, or a combination of children and grandparents living under the same roof. This represents a sharp reversal of long-term trends toward nuclear families. From 1940 to 1980, the percentage of Americans living in extended multi-generational households declined by half. The shift toward extended families comes from a number of different factors, including the trend of later marriage for many adults and the rising numbers of first and second-generation immigrants in the population.

Hispanic and Asian families, in particular, are used to living in multi-generational households. They often bring a history of values and extended family structures when they come to the U.S. Nonetheless, there are multiple cultural variations to what these extended families look like.

Asians are somewhat more likely to be living with elderly parents. Hispanics often share a home with aunts, uncles, or cousins. Immigrant families of European descent often consist of parents living with adult children.

For black families in the U.S., it is not unusual for children to be raised by their grandmother.

The recent increase in extended families is not so much a result of popularity as of necessity.

"We haven't seen anything like this since the Depression," says Frances Goldscheider, a sociologist who studies families at Brown University. "Overwhelmingly, it's the recession's effect on people's ability to maintain a house. You have foreclosures on one hand, and no jobs on the other. That's a pretty double whammy."[109]

When times are tough, it's just cheaper to live under one roof. This is a new type of "social security" – one that has nothing to do with big government or taxes, and everything to do with family.

In fact, one of the trends that we may see over the next few years is that some of the McMansions built over the past decade could become surprisingly useful for housing extended families.

Even for families not living together in the same household, there seems to be an increased reliance upon extended family members – particularly for taking care of children. "In Manhattan, people with kids have nannies; in Queens, we have grandparents," says Judy Markowitz.[110]

Eric Garland explains that the future of the American family will be living like the rest of the world. "Living together in multi-generational families is not a new concept in any way." It is a common living arrangement for families in many other countries. Garland says, "It's not new, but it will feel new to us."

Boomerang Kids and Parents

This shift toward extended families has affected all generations, but especially the elderly and the young. One in five adults aged 25-34 now live in a multi-generational household, along with one in five adults over the age of 65.[111]

There are more than 6 million elder Americans who live with their children. According to the Census Bureau, the number of people over the age of 65 living with their children has increased by over 50% between 2000 and 2010.

While institutionalized care was once seen as inevitable, the number of people over the age of 75 in nursing homes has been dropping since the mid-1980s. While part of this is accounted for by the increase in multi-generational families, other factors are also involved. Improvements in longevity mean that many people are able to delay going to a nursing home. Also, more families are having part-time help available to help provide caregiving services for their parents. [112]

At the other end of the spectrum, younger generations are more likely to return home after attending college. An article in the *Washington Post* tells a typical story: Daniel Sherrett is 28 and moved back to his mother's house in the Washington, D.C. area. He graduated from the Culinary Institute of America with a pile of student loans and now works as a waiter in a hotel restaurant. He lives in a converted garage apartment.

The arrangement is mutually beneficial for his family. Daniel looks after his autistic older brother, giving some relief for his mother. He also cooks for his mom and pays monthly rent. Someday, Daniel hopes to become a sommelier, but the competition is high.

Daniel's mother, Marie, has five sisters - all of whom also have adult children or nieces living with them. "But it's neat," she says. "You get companionship. In my case, you get someone to make you dinner." In this case, that someone happens to be a full-out chef.[113]

The Millennial generation faces the dual challenge of a weak labor market combined with competitive (and rising) educational standards. It's been over fifty years since so many young people in America were out of work. A recent study shows that one out of every three Americans between 18 and 29 are either choosing not to work or are looking for a paying job but cannot find one. There is a new twist to the contemporary job search: Millennials are often looking for the right opportunity and will not always accept an offer that does not fit their interests. Neil Howe, author of several books on generational patterns, says that Millennials are "more likely to take an unpaid internship, classes, or do free consulting – something that advances their goals."[114]

Meanwhile, Millennials generally seem to get along with their parents better than previous generations. In one 2007 survey, three-quarters of 13-24 year olds responded that time with family was their greatest source of happiness – even greater than time spent with friends or significant others. Half of the Millennials stay in daily contact with their parents, and almost all have weekly conversations.[115]

Households of One

Not only are families getting much bigger – some of them are getting much smaller. At the turn of the last century, only one in one hundred Americans lived alone. Today, the number is closer to one in ten.

The number of adults over the age of 65 living alone has decreased slightly over the past few years. According to a Pew Research Center study, older adults who live alone are more likely to report that they are not in good health and are more likely to feel sad, depressed, or lonely than their peers who are either living with a spouse or another family member. [116]

The percentage of the population living alone has grown slowly and consistently over the years. It is quite possible that this trend will reverse over the coming decade, however. The forces behind this will not only be economic, but social as well. In particular, there may be a greater range of available living arrangements.

Found Families

Found family can make life more pleasant and economically viable for everyone. It can run the gamut from a group of roommates sharing an apartment where real estate is expensive to neighbors who come to rely upon each other for help and support. *Found families* can sometimes be more temporary in nature, but they can be intentionally designed to help fulfill everyone's needs.

"Families are more diverse and the structure of them is more in flux," says Kelly Musick, a sociologist at Cornell University. "One of the things that's happened is people have a lot more leeway to design the families that work for them."[117]

I've developed a good friendship with a 10-year-old who is bright and fun to be around. He has a single mom, and until three years ago, had no male grownups in his family that he could connect with. Meanwhile, my wife Stephanie and I have no kids of our own, but we have a lot to share with others. So, he is now my little brother. We were originally matched through the local Big Brothers/Big Sisters program and are now a part of each other's found family.

Once or twice a month, we get together and do "guy stuff", like working in the yard or talking about comic books and our rock collections. It's a great relationship, and we each play an important part in each other's lives.

Here's another story: Stephanie Green is a young woman I met a month ago at a festival in southeastern Pennsylvania. Spread out on her blanket was her collection of books and crystals that she was selling to finance her move to Costa Rica. Behind her sunny disposition and mellow attitude, she was recovering from a failed marriage.

Trained as a massage therapist and vegan chef, Stephanie found her place in life a year ago when a friend of hers needed help with eldercare for his mother. Every day, Stephanie would visit with her and help with stretching and exercise. In the backyard, they started a garden together, where they grew flowers and herbs. On rainy days, they would paint together or play games. For Stephanie, being a caregiver is "more like being a friend than anything. In turn, you are taking care of yourself."

During a visit to Costa Rica, she received an offer to be a caregiver in residence for a Fellowship Community – an integrated environment where caregivers live and share housing and property with elders that they are assisting. Stephanie describes it as a cross between a "commune and a retirement community". This idea holds a great appeal for her progressive sensibilities.

Gender Blending

During periods of economic transition, gender roles can shift, change and adjust to meet family needs.

We've seen a number of changes in the role of women in the past century, starting with the role of "partner" at the end of the pioneer period when women followed their traditional role as mother, while also managing the household and the farm. After women were given political voice and access to voting because of the suffrage movement in the 1920s, they became further empowered economically as they went to the factories to work during World War II.

During the first baby boom, women went back home, rebranded as full-time mothers and consumers during a period of economic expansion. They went back to work in record numbers during the late 1960s and throughout the next three decades, the two-income household became the rule rather than the exception.

Women's roles in the media are changing as well. Women are now consistently portrayed as strong, independent, and assertive (if not aggressive) in television and film. These days, the "damsel in distress" is as likely to be a man as a woman.

One of the more telling instances of gender blending started to happen about ten years ago, when many of our friends started their families. At some point, my male colleagues stopped saying, "My wife is expecting" and instead started to proclaim, "*We're* pregnant." While this is a biological impossibility, it does tend to highlight changes in gender roles – perceived or real.

Over the past decade, we've seen a gradual shift in roles of men within the family, toward being more actively engaged in the process of nurturing and caring.

This seems to hold true for not only the U.S., but for Britain as well, where a recent survey by Aviva (an insurance company) revealed that the number of fathers in the UK staying home to look after their children has risen tenfold during the past ten years. "While both roles are equally valuable, nowadays it's quite likely that women will be heading off to the office while men are changing nappies and doing a school run," a spokeswoman from Aviva said.[118] The Aviva survey found that the woman is the main salary earner in 16% of families with dependent children.

In a single income household, the parent who stays home to care for offspring is typically the one without a paycheck. Given that the recent round of layoffs have hit male-dominated industries particularly hard (finance, construction, manufacturing), men have been pushed to make some significant adjustments.

Jeremy Adam Smith is a blogger and the author of *The Daddy Shift, a* collection of interviews with stay-at-home dads. He writes:

> *As men lost the ability to reliably support families on one income, families responded by diversifying. Men have developed emotional and interpersonal skills by*

taking care of children—since the mid-1990s, the number of hours dads spend with kids has nearly doubled—and women have gone to school and to work. In the eyes of many couples, equity between parents has moved from a nice ideal to an urgent matter of survival.[119]

There has also been a rise in men taking responsibility in caring for elderly parents. The Alzheimer's Association estimates that men make up nearly forty percent of family care providers – double the levels reported in a 1996 study.

Economic restructuring may promote a certain degree of role ambiguity out of necessity. During times of transition, it is important to make whatever contributions are needed by the family. Solidified gender roles work better during periods of economic expansion, when there is enough prosperity to support specialization.

Robert Prechter, editor of the *Elliot Wave Theorist* and founder of the discipline of socionomics, believes that there is a link between public mood, economic growth, and social trends. During a period of rising mood and economic growth, the media portrays men in film and television as "strong", and women as "beautiful". Heterosexual images with defined gender roles peak at the end of an economic expansion. During a falling transition, men are more frequently portrayed as "caring", while women are "liberated". [120]

Similarly, there is a very close correlation between performance of the stock market and annual conceptions in the U.S. As Prechter explains, "As people in general feel more energetic, confident, and happy, they have more children. Conversely, as people in general feel more sluggish, fearful, and unhappy, they have fewer children."[121] It is a simple explanation that appears accurate. Rising prosperity can support larger families for younger couples, while diminishing prosperity can result in delays in getting married and starting families.

Two Dads and Two Moms

Domestic roles can become even more interchangeable in gay and lesbian households.

In gay households the handling of chores often changes based upon the stage of the relationship. During the first year, partners typically share all of the chores. Later, however, routines become established and chores are assigned primarily on the basis of each partner's skill and schedule. For those duties for which each partner is capable, one partner may voluntarily "unlearn" skills to provide a complementary balance in the relationship.[122]

This contrasts somewhat with the division of labor in lesbian households, where couples are more likely to divide all housework equally, regardless of the stage of the relationship.[123] This seems to reflect an ethic of equality over that of specialization.

One of the biggest social changes over the past decade has been the increased acceptance of same-sex unions and marriages. Since Massachusetts first legalized same-sex marriage in 2004, a number of other coastal states have passed legislation approving marriage or domestic unions for gays and lesbians.

As same-sex couples gain greater legal and cultural acceptance, many of them are becoming more open to the idea of having families of their own. Between adoption, in vitro fertilization, and children from previous heterosexual relationships, it is estimated that there are roughly one million LSGBT (lesbian, gay, bi-sexual, and transgender couples) throughout the United States who are raising approximately two million children.[124]

About twenty percent of same-sex couples live with adopted children. This has more than doubled since 2000. "The trend line is absolutely straight up," says Adam Pertman, executive director for the Evan B. Donaldson Adoption Institute.[125]

Rob and Clay Calhoun have two adopted children – an eight-year-old daughter, Rainey, and a six-year-old son, Jimmy. Both children share the same biological mother.

"We're not moms, we're not heterosexual. We're not biological parents," says Rob. "We're totally equal and just as loving as female parents, as straight parents, and biological parents."

Rob also makes the point that "love makes a family, not biology, or gender."[126]

At least 21 states have granted two-parent adoptions to gay and lesbian families. Earlier this year, New Hampshire repealed its fifteen-year ban on adoption by gay couples. There are only two states, Utah and Mississippi, which have explicit legal barriers to gay adoption. For roughly half of the remaining states where same-sex marriage or domestic partnership is not recognized, there may still remain some complications to adoption process.

Marital Melting Pot

While gay marriage is still not universally recognized or accepted in the U.S., there is a much more firmly established trend toward mixed-race marriages. According to a report by the Pew Research Center, nearly 1 in 7 marriages are either interracial or interethnic. The intermarriage rate has more than doubled since the 1980s, and has increased six-fold since the 1960s. According to Daniel Lichter, a professor of sociology at Cornell University, "This indicates greater racial tolerance, a blurring of the racial divide in the U.S. In general, it's an optimistic report." [127]

Since 1980, blacks have nearly tripled their intermarriage rates, while whites have more than doubled theirs. Levels are relatively unchanged for Hispanics and Asians, who have consistently maintained higher levels of intermarriage in this country over the past few decades.

To a certain extent, this is a generational phenomenon. While the Boomer generation marched for peace and racial equality, their children have completed the shift from principle to practice. The Millennial generation, in particular, is the most racially diverse and accepting generation ever born in the U.S., with between 80-90% of respondents under the age of thirty reporting they believe interracial marriages are acceptable. This figure drops to about 30% for respondents over the age of 65. "People 65 and over grew up in the 30s, 40s, and 50s, when such things weren't acceptable or were illegal. That's an indicator of how things have changed," comments Jeffrey Passel, a lead researcher on the report. [128]

If interracial marriage is a sign of acceptance and equality, the statistics have clearly indicated that we are living in a changed world.

A Real Modern Family

Lauren Kim and Greg Lercel's family has many characteristics of the new demographics. They both have demanding jobs and after an extended period of graduate education, they had their first child in their thirties. Like many contemporary couples, they have a mixed-race marriage. As a physician, Lauren is the primary wage earner in the family.

Over the years, Lauren has maintained close ties to her Korean family. She provides monthly economic support to her parents, who live forty minutes away in Orange County. In exchange, her parents come and help them out from Wednesday to Friday each week. "That's sort of the arrangement" they have, Lauren says. During those days, her parents will shop and cook for the family, help out with light cleaning, and take care of the baby. On Friday night, they return to their own home.

On the remaining days of the week, Lauren and Greg have a babysitter during the day and equally share responsibilities at night.

In describing her extended family, Lauren shares that "I feel like we have a close relationship." Her parents are careful to respect their children's need for privacy, however, and they typically eat dinners separately. In the evenings they watch TV and fall asleep in the guest bedroom.

Earlier this year, Lauren and her husband purchased a second home, keeping in mind that her parents might be moving in with them on a full-time basis. The house still needs substantial renovations, but it has a guest cottage that is attached to the main house via a breezeway. The guest cottage has a bedroom, private bath, and its own kitchenette. Lauren and Greg asked her parents if they would like to move in with them on a full-time basis, but her parents declined the offer and wish to maintain separate households and a certain degree of independence for now.

This is a real "Modern Family" – one that embraces cultural roots and diversity, while adapting to the needs and preferences of everyone.

More Emerging Trends:

A Place for Grandma

One builder in California recently said that one-third of his buyers are looking for homes with "granny-flats", to provide housing for aging parents on-site, but with a greater degree of privacy.

Some areas, such as Santa Cruz, California, are changing their zoning rules to allow homeowners to convert garages, basements, and backyards into "accessory dwelling units." While the initial intention may be to provide nearby living space for mom and dad, they can later be converted into a rental unit to produce a supplemental source of income.

While it can sometimes cost hundreds of thousands of dollars to build a stand-alone home, a granny flat built on an existing property can be much less expensive. "The typical cost to build affordable housing (in Santa Cruz) is about $450,000," says architect Mark Primack. Meanwhile, he designed a garage conversion for a single mother that cost $36,000. Other layouts, such as stand-alone cottages, can be more expensive, but are still cheaper overall because they do not require the purchase of additional land. Roughly a third of all existing homes have the potential to build or convert existing space to accommodate a granny flat.

Granny flats provide other sources of flexibility. They could also be used as housing for boomerang kids, who want to stay close and be independent, yet cannot afford otherwise to live in the area.

Finally, granny flats can be used by the homeowners *themselves* when they are ready to downsize. When they are ready, they can move into the cottage, while their kids and grandchildren move into the main house.

Boarding Houses v. 2.0

People who find themselves with a big property that they can't sell and a large mortgage that they can no longer afford are getting creative with their cash flow. Renting out a room can also provide additional tax advantages, as well.

Unless there are friends or relatives who have an interest in renting, this means opening up the house to complete strangers. This type of *found family* is likely to boom if we see further economic pressures.

The details of these arrangements vary significantly. Some people will rent out multiple bedrooms with limited access to the remainder of the home, while other arrangements may be much smaller in scope but more inclusive.

Homesharing, Inc. is a non-profit in New Jersey that matches people who are looking for housing with homeowners who need either help or supplemental income. "What we try to create are new family relations," says executive director Renee Drell.   Depending on the situation, both parties can agree to reduced rents in exchange for help around the house.   "Most service arrangements are around errands, food preparation, and companionship," Drell explains.   Given that the costs of heating, taxes, and food have all increased, she has been seeing an increased demand for homesharing arrangements.

American Hikkikomori

A sustained economic downturn may result in an increase of hopelessness and a higher dropout rate. Ubiquitous access to virtual worlds combined with a lack of economic opportunities can create a generation of *Hikkikomori* – a Japanese term for people suffering from acute social withdrawal. In some instances, *Hikkikomori* are kids that grow into adults but never leave the house, preferring to immerse themselves in MMRPGS (massive multi-player role-playing games) and other pursuits of fantasy.

Three Parents or More

At the cutting edge of family, there may someday be more than two legitimate biological parents.   UK scientists have recently created "designer embryos" holding DNA from a man and two women.

Researchers at Newcastle University in the UK developed the process to overcome the problems of damaged mitochondrial DNA, which are passed exclusively from a mother's genes. In this process, a fertilized nucleus from an in-vitro fertilization is transplanted into an empty fertilized egg left from another (healthy) woman. This cell contains the healthy mitochondria, while transplanted nucleus contains the remaining genetic material.

While no children have been born of this process (yet), the opportunities for more permutations of the post-nuclear family are quite interesting...

Online Resources:

How To Rent Out a Room In Your Home

www.wikihow.com/Rent-a-Room-in-Your-House

Living With Needful Aging Parents

www.workingcaregiver.com/articles/safetytips/needfulagingparents

National Shared Housing Resource Center: clearinghouse of information sharing home sharing

www.nationalsharedhousing.org

Pew Research Center: surveys and data on social trends

www.pewresearch.org

Six Tips for Living with Boomerang Kids

www.bankrate.com/finance/financial-literacy/6-tips-for-living-with-boomerang-kids-1.aspx

Chapter 11

Retooling

Terry Bowles wasn't so sure he wanted to retire. What he *did* know was that he wanted to spend more time cruising through central Texas on his Honda Goldwing motorcycle.

Goldwings are built for touring and comfort. They have a smooth ride and can even carry enough cargo for an overnighter. Unlike riders of the Harley Davidson, who are trying to capture the iconic coolness of the classic American cycle, or the fans of the high-speed Italian Ducati bikes, Goldwing lovers just want to get there in style without losing touch with basic comforts.

Riding through Abilene on a long straight highway with a cup of coffee in hand can be a great way to spend a morning. The problem was the coffee. It was not a problem of drinking a hot 16-ounce latte while driving an 800-pound motorcycle one-handed at 70 mph; it was a problem of where to put the coffee in case you had to stop at a light or turn a corner. The Goldwing already had multiple coffee holders. The shortcoming with all of them was that coffee tended to spill whenever the motorcycle was in motion.

Terry set his engineering mind to work on this. The challenge was in understanding how to keep a hot cup of coffee vertical at all times. The solution was a gyroscopic coffee holder -- the weight at the bottom created a counter-balance to create the ideal holder for the perfect morning drink.

After finding a manufacturer to make the coffee holders, Terry turned his garage into a headquarters and did business by mail to other Goldwing riders. He then diversified into selling travel seat covers for the beloved cycles and soon thereafter started travelling to Goldwing conventions across the country. He was selling products to cover expenses and make a little extra money on the side.

This is a great example of turning passion into profits. Instead of "riding off into the sunset" on his Goldwing, Terry found a way to stay involved in his community of bikers and support himself and his wife, Gail. He never really thought of himself as an inventor or as someone in sales or marketing; it just came naturally for him.

Retirees across the country are itching to get things done. Many, while still "retired", are rejoining the workforce or starting their own businesses. Some are going back to school. Others are putting in long hours as volunteers for their church, neighborhood associations, or favorite cause.

Early retirement is becoming less about quitting work altogether and more about the ability do *something else*. It is a time when the children are grown, and if you are lucky, the mortgage is paid off. There may still be some expenses to pay, but it is a time to simplify and do things at your own pace.

In our traditional thinking about life paths, there are three distinct stages of approximately 25-35 years each. The first stage is that of education and learning to be productive. The second stage is filled to the brim with the dual tasks of working full-time while raising a family. The third stage is full retirement, with time spent in travel, grandchildren, and favorite hobbies.

An exceptional period of prosperity over the past few decades made extended retirement possible for members of the silent generation and the great generation.   During the Great Depression, life expectancy for men was only 67 years. Retirement at that time was a relatively short-term situation, often brought about by illness or disability. As such, it was rarely planned in any significant way. With the rise in big business during WWII and steady returns for the financial markets, employers encouraged long-term loyalty in the form of generous retirement benefits. Social Security was well funded by legions of young Boomers entering into the workplace.

The second and third stages become quite blurred in a zero-growth economy.   Extended periods of underemployment can exist in the second stage of life, while adding a job in retirement can be a financial necessity.

In a recent survey on retirement by the Employee Benefit Research Institute (EBRI), a record 74% of workers said they expected to work in retirement to supplement their finances.[129] The study concluded that Americans have accepted working longer as part of the "new normal."

Why Work?

Even the affluent are working longer, not out of necessity but out of a sense of purpose and self-fulfillment. Barclay's Wealth has labeled the current generation "nevertirees," as in never tiring, always going. Greg Davies, the head of the Behavioral Finance group at Barclays says that "for many, their working life is an important part of who they are – it is something from where we derive self-worth and value, and not just a necessary evil to be endured."[130]

Sarah Harper, a Professor of Gerontology at the University of Oxford feels that "people want to contribute; they want to be doing something. Work gives people status, and at an age when you're incredibly experienced, you may want to start a second career or even do something completely different from your previous professional life."[131]

Some may even choose to take these life changes in mid-career. Kathleen Wright had achieved a certain level of financial independence by her forties, and then quit her high-pressure job at a mining company to follow her passions to teach yoga and consult as a Vedic astrologer. In order to make the shift possible, it was necessary for her to downshift and simplify her life. She sold her house to build her personal portfolio and then moved from high-priced San Francisco to a smaller city. Today, Kathleen is vibrant and healthy in her mid-60s and enjoys a two-minute walking commute to her yoga studio on the first floor of the apartment building where she lives.

Joan Sharp has been a financial life planner for a decade now. For Joan, "It's about being true to who you are now. And it's about living now. What happens is that people who live true to who they are end up having more than enough resources." This shift started happening within the past few years. "When the crash happened, it really woke people up to who they were – just spending money mindlessly. It got people thinking about what was important to them."

As George Kinder describes it, "What happens is the process of life planning itself pulls people away from living a much more materialistic life and toward a life which is much more filled with spirit and filled with who they really are. "[132] As an early leader in the life planning movement, George asks three questions:

- "If you had all of the money that you needed, how would you change your life?"
- "If you only had 5-10 years to live, what would you do differently with your life?"
- "If you go to the doctor and he says you only have 24 hours to live… what did you miss? Who did you not get to be? What did you not get to do?"

Kinder has found that the third question goes to the deepest place for most people. And the answer to that question is where the life plan starts.

Our collective definition of retirement is changing in a very basic and fundamental way.

The "Money Pile"

Financial planners in the 1990s took the "money pile" approach to retirement funding. The working years of a couple were dedicated to building a pile of money. The planner's job was to estimate how much of a pile was needed for retirement. Once retirement was reached, the money was managed to provide growth and income. The goal here was to keep the pile from disappearing. Finally, the last task of the financial planner was to develop an estate plan for the equitable and efficient redistribution of the pile to children and loved ones at the end of retirement.

It was easy to manage the pile when money market funds earned 5% interest and Treasury Bonds paid 7%. The particularly daring could invest in junk bonds or emerging market debt and obtain double-digit yields.

When I did financial planning in the early 90s for a large national firm, the standard practice was to assume that a diversified portfolio could generate 8-11% per year in a combination of growth and income. At the time, this was considered to be reasonably conservative.

Financial planners still get nostalgic about the nineties and the turn of the millennium. It was a time in which the stock market generated average returns of 18% *per year*. The Federal government ran budget surpluses for two years. For a brief time, gasoline was cheaper than bottled water.

It has become a very different world. We've just completed the worst decade for stocks in over a century. Many retirees are no wealthier now than they were ten years ago, especially when the rising costs of living are considered. Seeking greater stability and income, they've fled into bonds, looking for safer, more reliable returns.

The New Bubble

This flight to quality has created a new bubble that has not yet burst – a bubble in bond prices.

Interest rates have fallen to such an extent that yields on many fixed-income investments such as CDs and savings accounts are now below the officially stated rate of inflation. In 2010, the popular inflation-indexed government bonds (TIPs) actually paid a *negative* yield. This meant that over time, investors were actually *paying* the government to hold onto their money. Interest rates have gotten so low that Charles Schwab and other brokerage firms have waived management fees on their money market accounts, because whatever modest yields they were earning on their cash accounts were completely eclipsed by their fees.

The surest sign of a bubble is behavior that *just doesn't make sense*. We've hit that point in bonds and what feels like a "sure thing" now is a recipe for systematic loss of purchasing power over the next decade.

This is all bad news for retirees who are looking to generate fixed income from their investments. Many retirees have been shifting their assets into lower quality, higher-yielding securities. Recently, the big trend has been moving money into emerging market bond funds, which currently have the same comparable yields as money market funds did two decades ago. In short, people are often taking higher risks with their nest egg to receive the same level of monthly income.

The timing couldn't be worse. From a historical perspective, we are close to the very low point of the interest rate cycle. According to the Foundation for the Study of Cycles, interest rates typically follow 60-year cycles, mirroring long-term trends in commodities prices. The current cycle of low interest rates may end sometime around 2013/2014. After the next year or two, look for interest rates to move substantially higher during the following two or three decades. Higher interest rates mean a higher cost of borrowing for governments and businesses – often putting additional pressure on weak institutions. Bond investors need to worry not only about the return *on* their money, but in some cases, the return *of* their money as well!

There is also the problem of retirement savings. According to the EBRI, the average Boomer between the ages of 55 and 65 had only $69,000 saved for retirement in 2009. Of greater concern is the average savings of people in the "retirement zone" between the ages 65 and 75. This group has only $56,000 worth of retirement savings per person. [133] It might not matter how much you can earn on investments if you only have one or two years of living expenses in your retirement account.

More Than Just Money

All of this suggests that it may become increasingly difficult to develop a long-term forecast of financial needs and income in retirement. This may provide a considerable amount of confusion for those who have already made a commitment to retire under the expectations that life will carry on as it has for the past few decades.

Retirement is not a number. It is not an age or an account balance. Retirement is a *place* where a person has a certain amount of choice in what to do next. There are a few shortcuts to get to that place. One is to simplify needs and reduce expenses. The second is to focus on wellness – a healthier body will get you there faster and with fewer interruptions. The third is to "tune-in" and open up to the opportunities that life gives you.

- **Simplifying** is all about managing the financial needs of retirement. By simplifying earlier in life, future retirees will have more cash flow to put toward savings, investments, or debt reduction. It will also prepare them for a lower-cost retirement, reducing needs for outside sources of income.

- **Wellness** is focused on improving the physical prospects of retirement. Being stronger and healthier will provide for better personal energy, better relationships, and the ability to enjoy life for a longer period of time.

- **Tuning-In** is ultimately about freedom. More specifically, it is about freedom from one's own self and previously established patterns and attitudes. It is well documented that people with a regular practice of meditation and/or prayer have a more positive outlook on life and experience less stress. Less structured yet contemplative activities such as walking, gardening, or crafts can have similar effects.

Retirees will feel many of the same pressures as recent college graduates, newlyweds, or families. There will be inflation and there will be uncertainty. The most successful at retirement will use many of the same strategies as other segments of society – some level of increased self-sufficiency combined with a sharing of collective resources. Many of these behavior patterns are outlined throughout this book – minimizing, sharing instead of buying, doing it yourself, homesteading, and finding new "gigs" or part-time work.

The quality of life in retirement is not reflected so much by what we have, but what we are able to do.

In the new world, retirement planning may start to look more like life coaching. If you know that you don't have enough saved for a conventional "full" retirement, but managed to pay off your mortgage and reduce your expenses, then the question might be "what now?"

More Emerging Trends:

Mortgages in Full Reverse

As retirees tap out savings, look for reverse mortgages to become more aggressively marketed. In a conventional mortgage, the resident of the house pays money to the bank, with the house held as collateral by the bank. After the last payment, the house is owned by the resident. In a reverse mortgage, the bank pays money back to the resident on a monthly basis, but owns the property at death.

Mike Wilson, owner of Integrity Financial Planning in Indiana explains further: "On the downside, the costs to take out a reverse mortgage are still relatively high, but they have been coming down. You need to plan to live in your home for many years in order to make the costs more affordable. You and your spouse have to be at least age 62 and basically have no or very little debt on your home."[134]

There is a further challenge associated with this trend. As reverse mortgages become more of a rule, and less of an exception, the size of inheritances may dwindle. This means that many Boomers have a diminished chance of inheriting the family nest.

From Fairways to Greenways

In the 90s and 00s there was an enormous boom in golf course development. Golf courses were considered a means by which developers could meet green space requirements while generating ongoing revenues and raising prices for the new homes built around them.

Jay Mottola, executive director of the Metropolitan Golf Association, says that the industry has lost 100 clubs a year for the past four years.

Retirees with less money and time might find themselves playing fewer rounds of golf. Some of those courses will go into redevelopment, while others will grow wild. Meanwhile, don't be surprised if a few of them get turned back into farmland.

The New Senior Classmen

A second career for many early retirees means personal re-tooling and re-branding. Just as it may take a factory several weeks or months of downtime to get refitted for a new production line, expect many retirees to spend their time learning new skills and trades.

Taking the trend further, many retirees are going back to school and living on campus. They fit right in with the existing students, wearing college sweatshirts and going to football games. The only difference is they are no longer drinking underage at tailgate parties.

More than 50 universities in the U.S. now offer "senior housing" either on the edge of campus or within short walking distance. The Village at Penn State University offers a continuum of care, including independent living, assisted living, and a skilled nursing facility. For those with the intellectual interest and the financial reserves, this may become an increasingly popular option.

Online Resources:

The Kinder Institute of Life Planning: Resources on finding a life planner

www.kinderinstute.org

Elderhostel Institute Network: Not-for-profit organization dedicated to providing education opportunities for older adults

www.roadscholar.org

New Retirement: Retirement planning calculators and tools

www.newretirement.com

Social Security Retirement Estimator: Get Social Security estimates based on your actual Social Security earnings record

www.ssa.gov/estimator/

Work Goes Strong: Blog for careers in retirement

www.work.lifegoesstrong.com/category/work/retirement-second-careers

TestQ: Career and motivational tests

www.testq.com/

Planned Seniorhood: Listings of college-affiliated active adult communities

www.plannedseniorhood.com/index.php/retirement-communities-2/college-affiliated

Chapter 12

Eldering

My grandmother was a small woman of French and Dutch descent who lived in a suburb outside of New York City. When I remember her home, I always recall the scents of coffee, sugared ham, and anisette toast. She was always up-to-date on current events and unusually willing to share her opinions. Despite her tiny frame, my grandmother would best be described as "feisty" and "determined".

When she reached her late seventies, she was diagnosed with breast cancer. Later, a CAT scan indicated that the cancer had spread to the part of her brain that was necessary for breathing. Her doctor advised that she could have the tumor surgically removed, but also said if her breathing stopped that she could potentially enter a coma.

My grandmother decided that it was not worth the cost or the trouble. She stayed at home and my mother moved in to take care of her. After a few months, she passed away in her sleep while my mom was preparing breakfast.

Before she died at the age of 82, Grandma said that she "could not have hoped for anything more" out of life. Not everyone gets to have this kind of choice. Not everyone has such a clear view of their life's end. For my grandmother, it was clear that personal dignity and independence were every bit as important as recovery from illness. Living well meant dying well.

For many, life is marked by struggle. Living takes effort; few things take greater effort than dealing with a terminal illness. We constantly refer to disease in terms of conflict, as in the "battle against heart disease" or the "fight against cancer." Dying is usually spoken of in terms of defeat and loss -- "she finally gave in after a three month bout of pneumonia."

Our medical system treats death as a failure and not as an inevitability. We all die. The awareness of this helps us to appreciate and become aware of every moment. Living with the fact that life is both temporary and fragile teaches us to express our gratitude without delay or hesitation. There is no need to wait for the perfect time to share ourselves with someone, because each moment *is* perfect. This spontaneity of sharing is such an important part of helping us to reconcile our past with our future.

My wife is a doctor at a nearby hospital and treats many people as they experience the end of their lives. She is an amazingly empathetic person and spends almost as much time caring for families as she does attending to patients.

Stephanie is trained as an infectious disease specialist, but many of her conversations serve as a sort of therapy for families as they are coming to terms with the impending passage of a loved one. It is a complicated mix of emotions that can range from love to grief, guilt to generosity, fatigue to relief. In working with families, she has noted that the family members who push the hardest for artificial medical support and intervention at the end of a patient's life tend to be the most distant – either physically or emotionally. They seek a delay of death not because the patient isn't ready, but because *they* are not ready.

We can keep people alive in a mechanical sense – through oxygen pumps, catheters, and intravenous infusions of nutrients. However, unless there is the hope of wellness or ability to communicate and recognize others, the potential for adequate emotional closure is minimal.

In many instances, it appears that we are attempting to use medical technologies to heal relationships. Dying well means living well.

One of the signs of maturity is the acceptance of the inevitability of dying. In doing so, death becomes somewhat less accidental – less a failure of the body and more of a soulful act. Transitioning is the process of preparation for a life passage that starts decades before the moment of death.

Starting the Transition

Transitioning can involve many small steps. These are all taken to provide a degree of closure and comfort to the surviving family and to prepare for the final transition.

On the administrative side, transitioning can start early by meeting with a legal advisor to draft the appropriate documents. At the most basic level, everyone should have a *simple will* to direct where their assets will go when they are gone. More people are choosing *advance healthcare directives* to express their wishes regarding life-prolonging medical treatments in the event of their incapacitation. They may also specify whether or not they wish to donate tissue or organs for others in need.

Finally, a *durable power of attorney* is a document that gives another person the ability to act on one's behalf regarding financial matters, such as paying bills or filing tax returns.

For all but the most basic documents, it is worthwhile consulting an attorney in your state of residence.

There also are other aspects of transitioning. On a purely physical level, transitioning sometimes resembles an act of simplifying things. This is a matter of going through "stuff" and separating the valuable and the sentimental from the clutter. As part of the transitioning process, some elders may choose to gift beloved heirlooms while still able to share the stories that belong to them. In telling family history and in sharing everything that needs to be said, it is possible for elders to leave legacies that are more memorable to younger generations.

One of the most overlooked aspects of the transitioning process is the importance of healing emotions and relationships. We all have a certain window of time to say the things that need to be said to the people that we care about the most. There is an opportunity to change how we will be remembered by others.

Finally, an important aspect of dying well is that of mentoring. This involves passing on skills and knowledge to others, so that they may also gain the wisdom of elders. This too, may involve sharing stories and history, to pass onward a sense of perspective that would otherwise be lost.

Dying is a truly personal process. The ideal outcome is to create a sense of closure not only for oneself, but for one's family as well. Everyone approaches transition differently. For some, the stages of dying may be more of a conscious process than for others.

My grandmother had a unique aspect to her own process of transitioning.

As teenagers, my sisters and I were always somewhat mystified by our grandmother's ever-growing collection of paper bags. She must have had several hundred of them – grocery store bags, gift bags, and shopping bags. All of these were used once or twice, then tucked and folded away. She was the wife of a successful banker and always well dressed, yet something of a "bag lady" in her own home.

She had convinced Mom that she needed all of those bags for some future event. She never gave any indication of what that might be. This, too, was a mystery. After she passed away, my parents spent a few months travelling back and forth to her house to settle the estate. Everything was packed away and passed on to relatives or given to charity. When the house was finally emptied, her last stack of books went into the final remaining paper bag.

As always, she was prepared and thinking ahead.

Elder Boom

With all of the gifts of longevity, we may find that an aging population is one that requires significant care and attention. While we can only hope that our elder population will be healthy and active, gains in longevity sometimes come with the cost of living with frailty. Living longer can sometimes mean living a more limited life – economically, mentally, and physically. Half of all people over the age of 65 have at least two chronic health conditions. Meanwhile, a study from the Center for Disease Control suggests that only one-fifth of Americans over the age of 65 stay physically active on a regular basis.

We may soon see the first generation in history to spend more time caring for elderly parents than for children.[135]

As Ted Fishman writes in *A Shock of Gray,* there are multiple factors that support the trend toward greater longevity. The population shift toward urban centers, improvements in education, public health, access to medical care, and reliable treatments for infections are the "main ingredients for a potion that foils early death and gives us the joys and sorrows of longer lives."

The same developments that enable a culture to become modernized will also change the demographic mixture of its population. Literacy has the effect of giving women a choice between focusing on career and family.   The pursuit of higher levels of education may delay the start of childbearing for some.   Moving a large segment of the population away from farms and into cities where people can be more productive and enjoy higher access to healthcare is another factor.

Eventually, this turns the demographic pyramid upside down, with a profusion of elders and a vastly diminished number of children.   The countries that experienced early modernization will be among the first to experience the full effects of the "age wave".

In the United States, the Institute on Aging reports that there are 17 million Americans between the ages of seventy-five and eighty-five. That figure is expected to double by 2050.   While centenarians are something of a rarity today, the population of Americans expected to live over the age of one hundred is expected to grow to 2.5 million during the same period.

Internationally, we see the same pattern, but in Japan and Europe the trends toward aging are even more pronounced.   By 2050, Japan is expected to see a net population decline as a result of couples choosing to have fewer children and tight immigration policies, while the ranks of the elderly are expected to swell to close to 40 percent of total population.

Europe's total population is expected to stabilize over the next few years, yet the proportion of its population over the age of sixty-five is anticipated to grow faster than any other age group.   By that time, close to three in ten Europeans will be older than sixty-five, one in six over seventy-five, and one in ten over the age of eighty.[136]

Economics are the driving factor behind almost all social change, and this will be particularly evident with regard to care of the elderly.   Just as the extended family disappeared during the boom decades of the 20th century, it is quite likely that extended family structures will return again during times of economic stress.

The costs are not insignificant.

A private room in a nursing home with skilled staff costs an average of $77,000 per year. At the other end of the spectrum, licensed personal care assistants that help with basic errands such as cooking, shopping, housekeeping cost an average of $18 per hour. Home health aides that help with basic activities such as bathing, dressing, and mobility are in a similar cost range.[137]

Many households cannot afford skilled nursing care for any sustained period of time. Given the choice of living in a nursing home or staying with family, most people would prefer to be under the care of their children. For families able to make the transition, the consolidation of households can greatly help cash flow by combining expenses such as rent and utilities under a single roof.

Choices will need to be made at all levels.

In the early 1960s, national defense was the government's largest expense. It comprised half of total spending. By 2011, our government reduced defense expenditures down to just 20% of the budget – even while fighting wars in Iraq and Afghanistan. According to the Congressional Budget Office (CBO), Medicare and Medicaid (combined) are now 23% of total expenditures, while Social Security is closer to 20%. This means that just under half of total government expenditures are targeted toward older Americans. When translated to dollars, government support to the average American over the age of 65 has grown to about $26,000 per year ($14,000 through Social Security and $12,000 through Medicare). [138]

These are enormous expenses, which everyone is motivated to sustain for as long as possible. People who have spent a lifetime paying into the social security system deserve to have some sort of benefit. The difficulty arrives when these benefits become such a significant portion of government outlays that they can no longer be ignored.

Regardless, the rising costs of unlimited healthcare for seniors will be an extremely divisive political battleground by the end of this decade. Tough decisions will need to be made in order to keep the system going. Medicare benefits may go into effect at a later age. Social Security benefits may be reduced for those who have other sources of income.

Sometime toward the second half of this decade, we may also see some type of healthcare rationing. Eventually, this may limit healthcare benefits for those who suffer severe cognitive disabilities or are under artificial life support. As a result, we may see considerable growth in palliative care as support grows for choosing a natural death.

The face of healthcare will be changed for decades, but we will also develop a new maturity toward our culture's acceptance of the inevitable transition from this life to the next.

A Wisdom Culture

Marty Knowlton, a co-founder of Elderhostel, said, "In many ways, elders are the dominant force in society. We control 75 percent of the wealth, and because we typically vote in greater numbers than younger people, we wield enormous political clout. Elders have the potential of influencing the political, economic, and cultural agendas of the future."[139]

In the best of all possible worlds, the inversion of the demographic pyramid could bring about a *wisdom culture*, marked by tolerance and maturity.

If we could imagine living with a strong generation of elders who have seen decades of success and folly, how would it change our perspective? Would it help us to maintain a long-term view that can encompass several generations? Or would it cause us to become increasingly conservative and more risk-averse?

Most Boomers are still in the Retooling phase of life. Culturally, we still value the image of the "aerobic grandparent" – busy, engaged, active, and involved. In many ways, this is simply an extension of middle age.

As more Boomers become elders, they may gradually feel a desire to spend less time *doing* and more time *being.* Action moves inward. Life becomes more meditative and the world becomes a quieter, less hurried place.

If done consciously, daily activities can take on an almost Zen-like level of perfection. If we live long enough, we can become experts at living. Reviewing the past enables us to understand the present. We can see some things more clearly with the benefit of hindsight. There is also wisdom to be gathered from seeing cycles and patterns over the course of many decades.

Some of these patterns involve personal pain and bodily decline. By tapping into these experiences, there is the potential to develop a great deal of empathy and compassion for others. While personal suffering will seldom be sought after, it may become accepted as an important part of spiritual cultivation. Driven by necessity and freed from the responsibilities of mid-life, there is a newfound freedom to explore faith and one's own soul.

These values are passed onto to family and community in the form of unconditional love and acceptance. Rabbi Zalman Schachter-Shalomi writes, "By example as well as by instruction (elders) can help family life become a training ground for contentment and inner satisfaction. They can model how to slow down our feverish pursuit of material possessions by embracing inner-directed values that stress unconditional love, self-acceptance, and service to others." [140]

When combined with the growth of multi-generational families, there is an opportunity for elders to cultivate younger generations and imbue them with a deeper sense of self and family. In today's busy world, many couples lack the time, patience, and attention to fully immerse themselves in parenting. Elders have all of these qualities in abundance. This does not suggest that elders will have sole responsibility for taking care of children – they may not necessarily have the energy or endurance. It merely says that child raising will become more of a shared experience.

The contributions of elders are significant, yet unmeasurable, in many ways. "While productive aging propagates the image of elders as active, engaged, and vital, ultimately it presents a rather weak and incomplete vision of life," says gerontologist Harry Moody. "By insisting on the productivity of the old, we put the last stage of life on the same level as the other stages. This sets up a power struggle over who can be the most productive, a competition that the old are doomed to lose."

"By celebrating efficiency and productivity, we abandon the moral and spiritual value of life's stage, stripping old age of meaning. What we need is a wider vision of late-life productivity that includes values such as altruism, citizenship, stewardship, creativity, and the search for faith. In short, we need a spiritual vision that recognizes the value of elder's non-economic contributions to society."[141]

These contributions to society may cultivate a more sustainable form of "social security", in which elders can enjoy a greater level of social acceptance and relevance, while their communities benefit from a higher level of social cohesion. It transcends transactions that are merely financial in exchange for a type of social currency. Eventually, we could redefine our concepts of aging successfully to include those that are able to move beyond material concerns in favor of leaving the world a better place.

More Emerging Trends:

Skype a Doctor

Digital monitoring of all sorts is becoming increasingly commonplace. Several companies are offering GPS tracking devices to monitor movement of Alzheimer's patients. These devices can help locate wanderers or they can send out an alert if the wearer moves beyond a pre-defined distance from home.

ADT offers a companion service that enables the wearer to push a button and receive assistance through an individually selected response plane (call an ambulance, next-door neighbor, or relative).

Over the course of the next decade, we are bound to see significant gains in telemedicine. The convergence of economic necessity, a shortage of doctors, and improvements in digital videoconferencing suggests that telemedicine may become mainstream within the next five to ten years.

"The technology has improved to the point where the experience of both the doctor and patient are close to the same as in-person visits, and in some cases better," says Dr. Kaveh Safavi, head of global health care for Cisco Systems.

Other starts ups, such as NuPhysicia, American Well, and Telehealthcare.com, are developing their own platforms to electronically connect patients with doctors.

Telemedicine encompasses a range of emerging technologies, from monitoring a patient's glucose levels remotely to electronic data sharing between physicians to sending patient reminders via cellphone.

Home Alone (with a robot)

Personalities, locations, and preferences suggest that not everyone will want to live in close contact with relatives. For many of these people, technological solutions may extend the frame for living independently.

Colin Angle, CEO of iRobot, feels that the opportunities for home health care robotics are enormous. iRobot is currently known as the manufacturer of Roomba vacuum cleaners and military robotics. As a starting point, the company developed a telepresence robot, called AVA, which can follow older people around their home. Anxious caregivers can have a remote video feed of their relatives, or enable a two-way video conference call if needed. If a nightly phone to an elderly parent is unanswered, the robot could navigate through the house to find them.

"A robot is never going to replace me" as a relative and caregiver of an aging parent, says Angle. "The part of robots is extending independent living."[142] With the help of robotic monitoring and assistance, people can live at home for a longer period of time.

Robotics will find their way into hospitals, as well. Panasonic has introduced a new robotic system that includes a bed that folds into a motorized wheelchair while carrying an adult. This could minimize the risks of injury during routine transfers. The company has also experimented with robot washstands that can wash a patient's hair from start to finish while 3D scanners and sixteen "fingers" gently massage the scalp. Panasonic has also developed an automated arm that can perform tasks in the kitchen.

In a world where there are more grownups than children, we can only hope that care and independence can be provided as long as possible. Automated care will eventually help to bridge the gap.

A Geriatric Peace

There is at least one potential positive to the rising costs of healthcare. Mark Haas, a political scientist at Duquesne University, feels that the greater costs of pensions and services could offer the world a "geriatric peace", as industrialized countries will eventually choose between maintaining military budgets and caring for their elderly.[143]

A more mature population is also less likely to engage in acts of aggression or war. Life is fragile and grandchildren are precious. With a smaller proportion of the population of fighting age (roughly 18-25), there will be less interest in fighting wars abroad – at least for the U.S., China, Japan, and most European countries.

Online Resources:

AARP: American Association of Retired People

www.aarp.org

Agis: AssistGuide Information Services. Eldercare and planning resources

www.agis.com

Compassion and Choices: Supports, educates, and advocates for choice of care at the end of life

www.compassionandchoices.org

Books:

From Ageing to Sageing: A Profound New Vision of Growing Older by Zalman Schachter-Shalomi

Coming Home: A practical and Compassionate Guide to Caring for a Dying Loved one by Deborah Duda

Chapter 13

A Path to Fulfillment

This book is about transforming a crisis into an opportunity through a process of re-imagining the future. As we explore the first steps of an emerging cultural shift, we have an amazing opportunity to solve the paradoxes of our age and grow toward a more balanced and sustainable system.   Like most ecologies, people do better when they are resilient.

There are means of achieving lasting prosperity without relying upon growth and consumption. If there is one thing that we can successfully predict, it is that people will continue to do everything that they can for themselves and their families.   Simplicity, self-reliance, and local community engagement cost nothing and can provide a more fulfilling life.

We have covered many different concepts over the past few chapters. Some of these changes are already happening in small ways. Many of them are still somewhat on the fringe. The best way to find trends early is to keep an eye out for changes on the periphery. Remember, the view is always better from the edge. When a dominant system fails, a counter-culture often ascends and captures the popular imagination. Positive social transformation *requires* radical elements. Conformity and stability will only possess value only after a more desirable state is reached. While most of the trends and movements reviewed in this book currently exist to a limited extent, the question remains whether they will gain enough participants to create a critical mass of social change.

One possible near-term future is that everything old is new again, as we go back to the basics and back to the real. For example, there are some interesting things happening with regard to the opposing forces of specialization and generalization. Times of economic expansion enable people to focus on developing deep experience within narrow skills sets. Specialization is a sign of progress and is economically more productive. The problem is that too much specialization makes people miserable and potentially vulnerable.

On the other hand, during periods of economic decline, people need to broaden their skill sets. This gives them more flexibility and choice and also enables them to do things themselves when they don't have the resources to hire someone else.

Having a broader range of life experience creates a fuller appreciation for the talents of others. In short, generalization creates a sense of choice, balance, *and* connection.

We'll all experience the future somewhat differently, and the ways that we adapt to change will be framed by our needs, values, and resources. What seems to be happening is that people are changing their lives in ways that are appropriate for their generation.

The members of the **Silent Generation** (born prior to 1945) may be among the last to live their entire lives in a traditionally modern way. They were among the primary beneficiaries of the decades-long growth cycle. Peace, stability, and expansion have been in place since their adolescence. Members of this generation may read the headlines, but relatively few adjustments may be required from this group.

The **Baby Boomers** (born between 1945 and 1960) have been agents of change since they were born. They were leaders of the counter-cultural movement in the 1960s and 1970s, and may be surprisingly quick to dial back their clocks and redefine our conceptions of aging. We will see this generation adapting through the trends of *retooling* and *re-extended families*. Many of them are involved in community-building efforts and the *Transition Town* movement. They may be the first generation of sages in an emerging *wisdom culture*. A changed set of social values will be this generation's legacy.

Generation X (1960-1980) may take self-sufficiency to an entirely new level. This generation is heavily involved in *gardening, homeschooling,* and the resurgence of *DIY*. These are the new wave of makers and tinkerers. Some Generation Xers are relearning old-fashioned skills to make their households more economically productive. Because Boomers are keeping their jobs longer, Gen Xers are often found starting their own businesses or looking for ways of supporting their families outside of the traditional workplace.

At this point, it appears that the **Millennials** (1981-2001) have reset their economic expectations and may be the first post-consumer generation. This could happen with a gradual shift away from *ownership* toward *access*. These are the new minimalists – younger adults are learning how to live with a minimum amount of "stuff." This group is learning how to *share, swap,* and *barter* better than any previous generation. Furthermore, what they do own is more likely to be stored electronically.

Millennials may increasingly become *digital nomads*, as their low-ownership lifestyle and Internet savvy enables them to work from anywhere, anytime. Their currency is reputation, and their goal is freedom and flexibility.

Change starts from within. Things will only get better after we start looking for the positives.

It is a bit like the punch line of an old Monty Python skit: "The floggings will continue until morale improves." We will see signs of improvement when people adjust their expectations and show confidence in the future.

More than anything else, the restructuring of the past decade has manifested itself as a collapse of trust at every level.   In the pursuit of "bigger, faster, more" of everything, we've lost contact with our basic values. We ask our government to fix things, but the government is too unresponsive and companies are too self-interested. In the end, it is just us.   We need to rebuild our collective sense of trust from the ground up.

The future gets brighter only when we can *imagine it.*

We have seven billion potential solutions wandering around the planet right now. Seven billion thinkers, doers, and dreamers.   On the whole, we are healthier, better educated, and more connected than ever before. There is a time to create a better tomorrow, and that time is now.

Appreciation

This book is based upon the ideas and contributions of many others.

My earliest "heroes of the future" include such thinkers as Alvin and Heidi Toffler, Pierre Teilhard de Chardin, Buckminster Fuller, and John Naisbitt. Reading a great book about the future can be like reading a good history. These writers were visionaries. The Tofflers' book *The Third Wave* anticipated many of the themes written about here.

Two significant figures who have been particularly instrumental in my development as a futurist include my graduate studies professors Peter Bishop and Oliver Markley, who continue to be both mentors and friends.

Throughout this book are interviews from practicing futurists, including Garry Golden, Andy Hines, Eric Garland, Jason Swanson, Heather Schlegel, Jared Weiner, and Joyce Goia. Many thanks for the conversations and input!

Many people have generously given me their time to be interviewed in person, by phone, or via e-mail. Their insights and stories have added readability and relevance. These include (but are not limited to) Haidy Dupuy, Bill Rees, Stephanie Green, Patrick Dugan, Hollis Thomases, Terry Bowles, Justin Caggiano, Joan Sharp, Helen Samson-Mullen, Doug Fine, Rachel Hoff, Tom Ferguson, Tom Wathen, Katie Fraser, Nick Kacher, Kathleen Wright, Lauren Kim, and Melanie Hardy.

Peggi Mitten is a long-time friend with an eye for detail and language who helped me improve upon early drafts.

My life coach, Suzanne Eder, helped make it all happen by encouraging me to follow my passion for writing.

Appreciation

A final thanks goes to family – my parents, for creating an environment of open inquiry and my wife Stephanie for her tolerance, patience, understanding, and support.

About the Author

James H. Lee is a Delaware-based financial advisor who engages in extensive research and analysis of emerging technologies and social trends. During his 20-year career, his system has identified numerous investment opportunities for his high net worth clients.

In recognition of his expertise, Lee has been interviewed by *The Wall Street Journal*, *Financial Planning*, and *Medical Economics*. He has also written for *The Futurist*, the *Journal of Futures Studies* and *Technological Forecasting and Social Change*.

Lee is a popular blogger for the World Future Society and has presented at their annual conference. He is also a member of the Association of Professional Futurists.

Lee graduated from the College of William & Mary with a B.A. in Economics and received a Master's degree in Studies of the Future from the University of Houston–Clear Lake. He is a Certified Financial Planner (CFP®), Chartered Financial Analyst (CFA), and a Chartered Market Technician (CMT).

Contact Info:

Email: lee.advisor@gmail.com

About the Author

Blogs: www.resilience-economics.com
www.wfs.org/content/james-h-lee

Twitter: @jhlinde

Endnotes

1 Paul Krugman, "The Big Zero," *NYTimes.com*, December 27, 2009,
http://www.nytimes.com/2009/12/28/opinion/28krugman.html?_r=1

2 Sharon Gannon, "Yoga and Money," in *What Comes after Money? Essays from Reality Sandwich on Transforming Currency & Community*, by Daniel Pinchbeck and Ken Jordan (Berkeley, CA: Evolver Editions, 2011), 131.

3 David Wiedemer, Robert A. Wiedemer, and Cindy S. Spitzer, *Aftershock: Protect Yourself and Profit in the next Global Financial Meltdown* (Hoboken, NJ: John Wiley & Sons, 2011), 6.

4 "Right Direction or Wrong Track," RassmussenReports.com, December 23, 2012, accessed December 23, 2012,
http://www.rasmussenreports.com/public_content/politics/mood_of_america/right_direction_or_wrong_track

5 "5% Say Congress Doing Good or Excellent Job," Rasmussen Reports, January 31, 2012,
http://www.rasmussenreports.com/public_content/archive/mood_of_america_archive/congressional_performance/5_say_congress_doing_good_or_excellent_job.

6 Ben Baden, "The Ranks of the Underemployed Continue to Grow," *US News and World Report*, October 19, 2011,
http://money.usnews.com/money/careers/articles/2011/10/19/the-ranks-of-the-underemployed-continue-to-grow.

7 Sherle R. Schwenninger and Samuel Sherraden, "The American Middle Class Under Stress," *New America Foundation*, April 2011,
http://growth.newamerica.net/sites/newamerica.net/files/policydocs/26-04-11%20Middle%20Class%20Under%20Stress.pdf.

8 Baden.

9 Ibid.

10 Mortimer B. Zuckerman, "Why the Jobs Situation Is Worse Than It Looks," *US News*, June 20, 2011,
http://www.usnews.com/opinion/mzuckerman/articles/2011/06/20/why-the-jobs-situation-is-worse-than-it-looks.

[11] Gregory White, "A New Guide To How The Middle Class Is Being Crushed By Rising Prices, Falling Wages, And Declining Security," Business Insider, April 29, 2011, http://www.businessinsider.com/america-middle-class-in-decline-2011-4.

[12] Ibid.

[13] Damon Vickers, *The Day after the Dollar Crashes: A Survival Guide for the Rise of the New World Order* (Hoboken, NJ: John Wiley & Sons, 2011), 7.

[14] "The World Top Incomes Database," PSE-Ecole D'économie De Paris, Groupe De Recherche Sur La Mondialisation Et Le Développement, January 11, 2012, http://g-mond.parisschoolofeconomics.eu/topincomes

[15] "CEO Pay: Feeding the 1%," 2011 Executive PayWatch, accessed March 12, 2012, http://archive.aflcio.org/corporatewatch/paywatch/.

[16] Ibid.

[17] Dave Cohen, "Decline of the Empire," Documenting The Demise Of The Middle Class, May 5, 2011, http://www.declineoftheempire.com/2011/05/documenting-the-demise-of-the-middle-class.html.

[18] Suzy Khimm, "Who Are the 1 Percent?," *Washington Post*, October 18, 2011, http://www.washingtonpost.com/blogs/ezra-klein/post/who-are-the-1-percenters/2011/10/06/gIQAn4JDQL_blog.html.

[19] Colin Dobrin, "Negative Equity: How Many Loans Are Underwater in Your State?," Credit Sesame Blog, November 17, 2011, http://www.creditsesame.com/blog/negative-equity-how-many-loans-are-underwater-in-your-state/.

[20] "U.S. National Debt Clock : Real Time," U.S. National Debt Clock : Real Time, accessed March 12, 2012, http://www.usdebtclock.org/.

[21] Terrance P. Jeffrey, "Obama Has Now Increased Debt More than All Presidents from George Washington Through George H.W. Bush Combined", CNS News, October 5, 2011, accessed March 12, 2012, http://cnsnews.com/news/article/obama-has-now-increased-debt-more-all-presidents-george-washington-through-george-hw.

[22] "One Giant Debt for Mankind: U.S. National Deficit Would Reach Almost to the Moon If Piled High in $5 Bills," Mail Online, May 24, 2011, http://www.dailymail.co.uk/news/article-1390090/One-giant-debt-mankind-U-S-national-deficit-reach-moon-piled-high-5-bills.html.

[23] "Corporate Profits After Tax (CP)." Federal Reserve Bank of Saint Louis. Accessed March 12, 2012. http://research.stlouisfed.org/fred2/series/CP.

[24] "FRB: H.6 Release--Money Stock and Debt Measures-March 8, 2012," Board of Governors of the Federal Reserve System, http://www.federalreserve.gov/releases/h6/current/.

[25] Bachman, Jess. "2012 Death and Taxes Poster." Death and Taxes 2012. It's the Government, in Six Square Feet. Accessed March 12, 2012. http://www.deathandtaxesposter.com/.

[26] Gus Lubin, "David Walker: The Greece Crisis Is Coming To America In 2012," Business Insider, February 16, 2010, http://articles.businessinsider.com/2010-02-16/markets/30021768_1_david-walker-fannie-and-freddie-chief-auditor.

[27] Ibid.

[28] "The Financial Condition and Fiscal Outlook of the U.S. Government," Peter G. Peterson Foundation, November 2011, http://www.pgpf.org/Chart-Archive/~/~/media/DCDF26EDAED447498CA7385FAA8A8FD9.pdf.

[29] David M. Walker, *Comeback America: Turning the Country around and Restoring Fiscal Responsibility* (New York: Random House Trade Pbks., 2010), 210.

[30] Peter G. Peterson Foundation.

[31] Kenneth R. Weiss, "Plague Of Plastic Chokes The Seas," *Los Angeles Times*, August 02, 2006, http://articles.latimes.com/2006/aug/02/local/me-ocean2.

[32] Ibid.

[33] "Conservation and Research," New England Aquarium/The Consortium for Wildlife Bycatch Reduction, accessed March 12, 2012, http://www.neaq.org/conservation_and_research/projects/project_pages/consortium_for_wildlife_ bycatch_reduction.php.

[34] Kenneth R. Weiss, "A Primeval Tide of Toxins," *Los Angeles Times*, July 30, 2006, http://articles.latimes.com/2006/jul/30/local/me-ocean30.

[35] Lauren Morello, "Oceans Turn More Acidic Than Last 800,000 Years: Scientific American," Science News, Articles and Information, February 22, 2010, http://www.scientificamerican.com/article.cfm?id=acidic-oceans.

[36] Weiss.

[37] "Lake Mead Could Be Dry by 2021," Scripps Institution of Oceanography, February 12, 2008, http://scrippsnews.ucsd.edu/Releases/?releaseID=876.

[38] Paul Quinlan, "Lake Mead's Water Level Plunges as 11-Year Drought Lingers," *The New York Times*, August 13, 2010,

http://www.nytimes.com/gwire/2010/08/12/12greenwire-lake-meads-water-level-plunges-as-11-year-drou-29594.html.

[39] Charles Laurence, "US Farmers Fear the Return of the Dust Bowl," *The Telegraph*, March 7, 2011, http://www.telegraph.co.uk/earth/8359076/US-farmers-fear-the-return-of-the-Dust-Bowl.html.

[40] Ibid.

[41] Lester Brown, "Plan B Updates," Earth Policy Institute, August 6, 2002, http://www.earth-policy.org/plan_b_updates/2002/update15

[42] Carmen Revenga, "Will There Be Enough Water?," World Resources Institute, October 2000, http://earthtrends.wri.org/pdf_library/feature/wat_fea_scarcity.pdf.

[43] Sheila Moorcroft, "Water, the Oil of the 21st Century - Pressure Is Rising," Innovation Management, November 2, 2011, http://www.innovationmanagement.se/2011/11/02/water-pressures-rising/.

[44] "Water - Adapting to a New Normal," in The Post Carbon Reader: Managing the 21st Century's Sustainability Crises, ed. Richard Heinberg and Daniel Lerch, by Sandra Postel (Healdsburg, CA: Watershed Media, 2010), 82.

[45] Brown.

[46] Brown.

[47] "New Projection Shows Global Food Demand Doubling by 2050," PhysOrg.com, November 21, 2011, http://www.physorg.com/news/2011-11-global-food-demand.html.

[48] Kevin Drum, "Crude Awakening," *Washington Monthly*, June 2005, http://www.washingtonmonthly.com/features/2005/0506.drum.html.

[49] Alma Alsharif and Reem Shamseddine, "Aramco to Inject CO2 into Biggest Oilfield by 2012," Reuters, February 15, 2010, http://www.reuters.com/article/2010/02/15/saudi-co2-ghawar-idUSLDE61E0TW20100215?rpc=401&feedType=RSS&feedName=rbssEnergyNews&rpc=401.

[50] Drum.

[51] "International Energy Statistics," U.S. Energy Information Administration (EIA), accessed March 12, 2012, http://www.eia.gov/cfapps/ipdbproject/iedindex3.cfm?tid=5.

[52] Adam Grubb, "Peak Oil Primer," Post Carbon Institute, accessed March 12, 2012, http://www.energybulletin.net/primer.php.

[53] "2010 Key World Energy Statistics," International Energy Agency, 2010, http://www.iea.org/textbase/nppdf/free/2010/key_stats_2010.pdf.

[54] "2011 World Energy Outlook Executive Summary," International Energy Agency, 2011, http://www.iea.org/weo/docs/weo2011/executive_summary.pdf.

[55] United States, Joint Forces Command, *2010 Joint Operating Environment (JOE)*, 2010, http://www.jfcom.mil/newslink/storyarchive/2010/JOE_2010_o.pdf.

[56] Bill McKibben, *Eaarth: Making a Life on a Tough New Planet* (New York: St. Martin's Griffin, 2011), 2.

[57] "1st Commercial Ship Sails through Northwest Passage," CBCnews, November 28, 2008, http://www.cbc.ca/news/canada/north/story/2008/11/28/nwest-vessel.html.

[58] "The Curse of Carbon," The Economist, December 30, 2008, http://www.economist.com/node/12798428.

[59] Jonathan Amos, "Methane Ices Pose Climate Puzzle," *BBC News*, December 13, 2006, http://news.bbc.co.uk/2/hi/science/nature/6166011.stm.

[60] U.S. Global Change Research Program, *Global Climate Change Impacts in the United States*, ed. Thomas R. Karl, Jerry M. Melillo, and Thomas C. Peterson (New York: Cambridge University Press, 2009), accessed March 12, 2012, http://downloads.globalchange.gov/usimpacts/pdfs/climate-impacts-report.pdf.

[61] Jeff Goodell, "The Prophet," *Rolling Stone*, October 31, 2007.

[62] Z. Chen et al., *A Report of Working Group I of the Intergovernmental Panel on Climate Change*, report, ed. S. Solomon, D. Qin, and M. Manning (New York: Cambridge University Press, 2007), http://www.ipcc.ch/pdf/assessment-report/ar4/wg1/ar4-wg1-spm.pdf.

[63] John Renesch, *The Great Growing Up: Being Responsible for Humanity's Future* (Prescott, AZ: Hohm Press, 2011), 11.

[64] "Little Boxes, Little Boxes," Free Enterprise Land, accessed March 13, 2012, http://www.freeenterpriseland.com/BOOK/LITTLEBOXES.html.

[65] Renesch, 44.

[66] Robert Samuelson, "A Darker Future for United States," *Newsweek*, November 10, 2008.

[67] "Credit Card Debt Elimination and Facts About Debt In America," Spend On Life, accessed March 13, 2012, http://www.spendonlife.com/content/CreditCardDebtEliminationAndFactsAboutDebt InAmerica-1-223-3.ashx.

[68] Vickie Elmer, "Refinancing While Underwater," *New York Times*, September 1, 2011

[69] Donella H. Meadows, Jørgen Randers, and Dennis L. Meadows, *The Limits to Growth: The 30-year Update* (White River Junction, Vt: Chelsea Green Pub., 2004), 54.

[70] Global Footprint Network, accessed March 13, 2012, http://www.footprintnetwork.org/en/index.php/GFN/page/footprint_for_nations/.

[71] "The Dent Method," H. Dent, accessed March 13, 2012, http://www.mossohsdent.com/The_Dent_Methods.

[72] William Strauss and Neil Howe, The Fourth Turning: An American Prophecy (New York: Broadway Books, 1997).

[73] Eamon O'Hara, "Focus on Carbon 'Missing the Point'" BBC News, July 30, 2007, http://news.bbc.co.uk/2/hi/science/nature/6922065.stm.

[74] Andy Hines, "Changing Values and "Enoughness" Suggest Economic Stimulus Won't Work," Hinesight, September 22, 2011, http://www.andyhinesight.com/values/changing-values-and-enoughness-suggest-economic-stimulus-wont-work/.

[75] Mark Frauenfelder, "The Courage to Screw Up: Why DIY Is Good for You," The Huffington Post, November 17, 2011, http://www.huffingtonpost.com/mark-frauenfelder/home-diy-the-courage-to-s_b_589371.html.

[76] "Mark Frauenfelder," Colbert Nation, June 8, 2011, http://www.colbertnation.com/the-colbert-report-videos/311944/june-08-2010/mark-frauenfelder.

[77] Faythe Levine and Cortney Heimerl, Handmade Nation: The Rise of DIY, Art, Craft, and Design (New York: Princeton Architectural Press, 2008), 123.

[78] Ibid.

[79] Betsy Greer, "Knitting for Good!," Craftivism, accessed March 13, 2012, http://craftivism.com/book.html.

[80] Justin Lahart, "Tinkering Makes a Comeback Amid Crisis," *Wall Street Journal*, November 13, 2011.

[81] "Doug on the Tonight Show," DougFine.com, accessed March 14, 2012, http://www.dougfine.com/doug-on-the-tonight-show/.

[82] "Industry Statistics and Projected Growth," Organic Trade Association, accessed March 14, 2012, http://www.ota.com/organic/mt/business.html.

[83] Educating Our Children: The Evolution of Home Schooling, dir. Maggie Kerkman, Fox News, February 09, 2011, http://www.foxnews.com/us/2011/02/09/educating-children-evolution-home-schooling/.

[84] Lawrence Rudner, "Scholastic Achievement and Demographic Characteristics of Home School Students in 1998," *Education Policy Analysis Archives*, March 23, 1999.

[85] Jennifer Dukes-Lee, comment on "When Your Work Isn't Working," *The High Calling* (web log), June 15, 2010, http://www.thehighcalling.org/8725/when-your-work-isnt-working.

[86] Oliver Strand and Joe DiStefano, "Chefs Look for Wild Ingredients Nobody Else Has," *The New York Times*, November 23, 2010.

[87] Cohousing.org, accessed March 14, 2012, www.cohousing.org.

[88] Danielle Sacks, "The Sharing Economy," FastCompany.com, April 18, 2011, http://www.fastcompany.com/magazine/155/the-sharing-economy.html.

[89] Latitude and Shareable Magazine, "The New Sharing Economy," March 14, 2012, , http://latdsurvey.net/pdf/Sharing.pdf.

[90] Ibid.

[91] Alex Pasternak, "SharedEarth.com: A Landshare Grapevine Linking Gardeners With Gardens," Treehugger, April 29, 2010, http://www.treehugger.com/files/2010/04/sharedearthcom-landshare-grapevine-linking-gardens-gardeners.php.

[92] Paul Murray and Charlotte Murray, "Couchsurfing," Paul and Charlotte's World Travels (web log), accessed March 14, 2012, http://paulandcharlottesworldtravels.blogspot.com/p/couchsurfing.html.

[93] Renesch, 40.

[94] Anne Thomas, "A Letter from Sendai," Ode Magazine, March 14, 2011, www.odemagazine.com/blogs/readers_blog/24755/a_letter_from_sendai.

[95] Josh Voorhees, "U.S. Vehicle Fleet Shrank 2% Last Year, Biggest Decline in Decades," *New York Times*, January 6, 2010, www.nytimes.com/gwire/2010/01/06/06greenwire-us-vehicle-fleet-shrank-2-last-year-biggest-de-80794.html.

[96] U.S., Department of Transportation, *Traffic Volume Trends,* March 2011, www.fhwa.dot.gov/ohim/tvtw/11martvt/11martvt.pdf.

[97] Lester R. Brown, "U.S. Car Fleet Shrank by Four Million in 2009 - After a Century of Growth, U.S. Fleet Entering Era of Decline," EarthPolicy Institute, January 5, 2010, http://www.earth-policy.org/index.php?/plan_b_updates/2010/update87.

[98] Les Christie, "Say Goodbye to the McMansion," CNNMoney, August 26, 2010, http://money.cnn.com/2010/08/26/real_estate/the_typical_homeowner/index.htm.

[99] Lindsay William-Ross, "Old Fashioned Food Swapping Gets Modern Upgrade With New Community-Building Website," LAist, November 22, 2011, http://laist.com/2011/11/22/food_swapping_goes_tech_with_commun.php.

[100] Helen Jones, "Deli Dollar Offers Route to Business Funding," *The Independent*, February 17, 1999, http://www.independent.co.uk/news/business/delidollar-offers-route-to-business-funding-1071370.html.

[101] Stella Osojoros, "Time Banking in Santa Fe," in *What Comes after Money?: Essays from Reality Sandwich on Transforming Currency & Community*, ed. Daniel Pinchbeck and Ken Jordan (Berkeley, CA: Evolver Editions, 2011), 119.

[102] *The Case for Collaborative Consumption*, perf. Rachel Botsman, TED, December 2010, http://www.ted.com/talks/lang/en/rachel_botsman_the_case_for_collaborative_consumption.html.

[103] David M. Autor, *The Polarization of Job Opportunities in the U.S. Labor Market*, National Bureau of Economic Research, MIT Department of Economics, April 2010, The Polarization of Job Opportunities in the U.S. Labor Market. http://economics.mit.edu/files/6343.

[104] Cali Ressler and Jody Thompson, *Why Work Sucks and How to Fix It: The Results-only Revolution* (New York: Portfolio/Penguin, 2011), 6.

[105] Ibid, 82.

[106] "Eyes Wide Open Wallet Half Shut," Ogilvy, accessed March 14, 2012, http://assets.ogilvy.com/truffles_email/eyeswideopen_press/Eyes_wideshut.pdf.

[107] "Americans' Job Satisfaction Falls to Record Low," *USA Today*, January 6, 2010, http://www.usatoday.com/money/workplace/2010-01-05-job-satisfaction-use_N.htm.

[108] Bruce Feiler, "What 'Modern Family' Says about Modern Families," *New York Times*, January 21, 2011, http://www.nytimes.com/2011/01/23/fashion/23THISLIFE.html?_r=1&pagewanted=all.

[109] Carol Morello and Ted Mellnik, "Recession Pushes More in D.C. Area to Live with Relatives," Washington Post, August 18, 2011, http://www.washingtonpost.com/local/recession-pushes-more-in-dc-area-to-live-with-relatives/2011/08/17/gIQANHTaMJ_story.html?nav=emailpage

[110] Joel Kotkin, "All in the Family," Forbes, April 13, 2010, http://www.forbes.com/2010/04/12/family-immigration-millennials-opinions-columnists-joel-kotkin.html?boxes=opinionschannellatest.

[111] "The Return of the Multi-Generational Family Household," Pew Social & Demographic Trends, accessed March 18, 2010, http://www.pewsocialtrends.org/2010/03/18/the-return-of-the-multi-generational-family-household/.

[112] Kotkin.

[113] Michael Temchine, "Re-Extended Families," The Washington Post, accessed June 6, 2012. http://www.washingtonpost.com/local/re-extended-families/2011/08/17/gIQA4QxZMJ_gallery.html#photo=1

[114] Jenna Fisher, "Millennials Keep Their Chins up despite High Unemployment in Economic Downturn," The Christian Science Monitor, April 23, 2010, http://www.csmonitor.com/Business/2010/0423/Millennials-keep-their-chins-up-despite-high-unemployment-in-economic-downturn.

[115] Kotkin.

[116] Pew Research Center.

[117] Sharon Jayson, "What Does a 'Family' Look Like These Days?," USA Today, November 25, 2010, http://www.usatoday.com/yourlife/sex-relationships/marriage/2010-11-18-pew18_ST_N.htm.

[118] "Tenfold Rise in Stay-At-Home Fathers in 10 Years," The Guardian, April 6, 2010, http://www.guardian.co.uk/lifeandstyle/2010/apr/07/rise-stay-at-home-fathers-study

[119] Jeremy A. Smith, "Freer, Messier, Happier," YES! Magazine, December 2, 2010, http://www.yesmagazine.org/issues/what-happy-families-know/freer-messier-happier.

[120] Robert R. Prechter, The Wave Principle of Human Social Behavior and the New Science of Socionomics (Gainesville, GA: Elliott Wave International/New Classics Library, 2002), 233.

[121] Robert R. Prechter, Pioneering Studies in Socionomics (Gainesville, GA: New Classics Library, 2003), 60.

[122] David P. McWhirter and Andrew M. Mattison, *The Male Couple: How Relationships Develop* (Englewood Cliffs, NJ: Prentice-Hall, 1984).

[123] L.A. Kurdek, "The Allocation of Household Labor in Gay, Lesbian, and Heterosexual Couples", *The Journal of Social Issues, vol. 49, 127-139. 1993.*

[124] Nicola Menzie, "NYC Gay Marriage to Boost Same-Sex Adoptions?," Christian Post, January 13, 2011, http://www.christianpost.com/news/new-york-gay-marriage-law-to-boost-same-sex-adoptions-52223.

[125] Sabrine Tavernise, "Adoptions by Gay Couples Rise, Despite Barriers," *New York Times,* June 14, 2011, http://www.nytimes.com/2011/06/14/us/14adoption.html?pagewanted=all.

[126] Taylor Gandossy, "Gay Adoption: A New Take on the American Family," *CNN.com*, June 25, 2007, http://articles.cnn.com/2007-06-25/us/gay.adoption_1_gay-adoption-straight-parents-williams-institute?_s=PM:US.

[127] Husan Haq, "Interracial Marriage: More than Double the Rate in the 1980's", *Christian Science Monitor*, June 4, 2010.

[128] Ibid.

[129] Ruth Helman and Craig Copeland, "The 2011 Retirement Confidence Survey: Confidence Drops," *EBRI Issue Brief*, March 2011, http://www.ebri.org/pdf/briefspdf/EBRI_03-2011_No355_RCS-2011.pdf.

[130] "Age Won't Keep Us From Work, Say a New Breed of 'Nevertiree' Wealthy," Barclays Wealth Insights, September 27, 2010, http://www.barclayswealth.com/insights/age-wont-keep-us-from-work-say-a-new-breed-of-nevertiree-wealthy.htm.

[131] Ibid.

[132] George Kinder, "Life Planning Is Financial Planning Done Right," Kinder Institute, April 25, 2008, http://www.kinderinstitute.com/media/video-clips.htm.

[133] Helman and Copeland.

[134] Mike Wilson, "Where Retirees Can Shop for High Income in a Period of Low Interest Rates," http://integrityplanner.com/images/Where_Retirees_Can_Shop_for_High_Income_in_a_Period_of_Low_Interest_Rates.pdf

[135] Ted C. Fishman, *Shock of Gray: The Aging of the World's Population and How It Pits Young Against Old, Child against Parent, Worker against Boss, Company against Rival, and Nation against Nation* (New York: Scribner, 2010).

[136] *Preparing for an Aging World: The Case for Cross-national Research* (Washington, D.C.: National Academy Press, 2001), 33.

[137] "Executive Summary - Genworth 2011 Cost of Care Survey," Genworth Financial, April 27, 2011, http://www.genworth.com/content/etc/medialib/genworth_v2/pdf/ltc_cost_of_car e.Par.85518.File.dat/Executive%20Summary_gnw.pdf.

[138] Robert J. Samuelson, "Why Are We in This Debt Fix? It's the Elderly, Stupid.," *Washington Post*, July 28, 2011, http://www.washingtonpost.com/opinions/why-are-we-in-this-debt-fix-its-the-elderly-stupid/2011/07/28/gIQA08Ltfl_story.html.

[139] Zalman Schachter-Shalomi and Ronald Miller, *From Age-ing to Sage-ing* (New York: Grand Central Publishing, 1997), 229.

[140] Ibid, 221.

[141] Ibid, 214.

[142] Mariette DiChristina, "TED MED: Grandma's Little Robot Helper," Scientific American Blog Network, October 30, 2009, http://blogs.scientificamerican.com/observations/2009/10/30/ted-med-grandmas-little-robot-helper/.

[143] Fishman.